Exploring Tropical Cyclones
GIS Investigations for the Earth Sciences

Michelle K. Hall-Wallace
C. Scott Walker
Larry P. Kendall
Christian J. Schaller

The University of Arizona

Australia • Canada • Mexico • Singapore • Spain • United Kingdom • United States

Editor: *Keith Dodson*
Assistant Editor: *Carol Ann Benedict*
Editorial Assistant: *Heidi Blobaum*
Marketing Manager: *Ann Caven*
Advertising Project Manager: *Laura Hubrich*
Project Manager, Editorial Production: *Tom Novack*

Print/Media Buyer: *Kristine Waller*
Permissions Editor: *Sue Ewing*
Cover Designer: *Denise Davidson*
Cover Image: *Digital Vision*
Cover Printer: *Transcontinental*
Printer: *Transcontinental*

COPYRIGHT © 2003 Brooks/Cole, a division of Thomson Learning Inc. Thomson Learning™ is a trademark used herein under license.

ALL RIGHTS RESERVED. No part of this work covered by the copyright hereon may be reproduced or used in any form or by any means—graphic, electronic, or mechanical, including but not limited to photocopying, recording, taping, Web distribution, information networks, or information storage and retrieval systems—without the written permission of the publisher.

Printed in Canada

2 3 4 5 6 7   06 05 04 03

For more information about our products, contact us at:
**Thomson Learning Academic Resource Center
1-800-423-0563**

For permission to use material from this text, contact us by:
Phone: 1-800-730-2214   **Fax:** 1-800-730-2215
Web: http://www.thomsonrights.com

All products used herein are used for identification purposes only and may be trademarks or registered trademarks of their respective owners.

Maps and screen shots include data from ESRI Data and Maps. Data Copyright © ESRI 2002.

ESRI and ArcView are registered trademarks in the United States and are either trademarks or registered trademarks in all other countries in which they are used. The ArcView logo is a trademark of Environmental Systems Research Institute, Inc.

Earth surface image data used in the Earth globe illustrations on pages 15–17 are © 2001, TerraMetrics, Inc. Used with permission.

Development of these materials was supported, in part, by the National Science Foundation under Grant No. DUE-9555205. Any opinions, findings, and conclusions or recommendations expressed in these materials are those of the authors and do not necessarily reflect the views of the National Science Foundation.

Library of Congress Control Number: 2002102236

ISBN: 0-534-39147-8

**Brooks/Cole—Thomson Learning
511 Forest Lodge Road
Pacific Grove, CA 93950
USA**

Asia
Thomson Learning
5 Shenton Way #01-01
UIC Building
Singapore 068808

Australia
Nelson Thomson Learning
102 Dodds Street
South Melbourne, Victoria 3205
Australia

Canada
Nelson Thomson Learning
1120 Birchmount Road
Toronto, Ontario M1K 5G4
Canada

Europe/Middle East/Africa
Thomson Learning
High Holborn House
50/51 Bedford Row
London WC1R 4LR
United Kingdom

Latin America
Thomson Learning
Seneca, 53
Colonia Polanco
11560 Mexico D.F.
Mexico

Spain
Paraninfo Thomson Learning
Calle/Magallanes, 25
28015 Madrid, Spain

Acknowledgments

The authors wish to thank the many students, teachers, and scientists who, through their use of these materials, provided critical reviews and helped us develop insight into how GIS can be most effectively used as a learning and teaching tool.

A significant number of people contributed directly or indirectly to the development of this module, but a few were especially notable. Particular thanks go to Terry Wallace, Joshua Hall, Christine Donovan, Tekla Cook, and Richard Spitzer who tested the investigations multiple times in their classrooms and tirelessly worked with us to improve their content and design. Douglas Yarger, Robert Butler, Terry Wallace, and Joseph Watkins provided content reviews while Carla McAuliffe and Jo Dodds performed insightful reviews of the pedagogy, design, and content level of the materials. We also appreciate the considerable efforts of our student assistants Marie Renwald, Anne Kramer Huth, Tammy Baldwin, Sara McNamara, Megan Sayles, and Christine Hallman.

We are indebted to the numerous scientists who took the time to learn about our project and share critical research data or expertise that added greatly to the quality of the investigations. Finally, we are grateful to the agencies and individuals that have given us permission to include their outstanding illustrations and photos.

The SAGUARO Project
Michelle K. Hall-Wallace, Director
Department of Geosciences
The University of Arizona
1040 E Fourth Street • Tucson, AZ 85721-0077
saguaro@geo.arizona.edu

Science And GIS Unlocking Analysis & Research Opportunities
http://saguaro.geo.arizona.edu

ESRI Software License Agreement

This is a license agreement and not an agreement for sale. This license agreement (Agreement) is between the end user (Licensee) and Environmental Systems Research Institute, Inc. (ESRI), and gives Licensee certain limited rights to use the proprietary ESRI® desktop software and software updates, sample data, online and/or hard-copy documentation and user guides, including updates thereto, and software keycode or hardware key, as applicable (hereinafter referred to as "Software, Data, and Related Materials"). All rights not specifically granted in this Agreement are reserved to ESRI.

Reservation of Ownership and Grant of License: ESRI and its third party licensor(s) retain exclusive rights, title, and ownership of the copy of the Software, Data, and Related Materials licensed under this Agreement and, hereby, grants to Licensee a personal, nonexclusive, nontransferable license to use the Software, Data, and Related Materials based on the terms and conditions of this Agreement. From the date of receipt, Licensee agrees to use reasonable effort to protect the Software, Data, and Related Materials from unauthorized use, reproduction, distribution, or publication.

Copyright: The Software, Data, and Related Materials are owned by ESRI and its third party licensor(s) and are protected by United States copyright laws and applicable international laws, treaties, and/or conventions. Licensee agrees not to export the Software, Data, and Related Materials into a country that does not have copyright laws that will protect ESRI's proprietary rights. Licensee may claim copyright ownership in the Simple Macro Language (SML™) macros, AutoLISP® scripts, AtlasWare™ scripts, and/or Avenue™ scripts developed by Licensee using the respective macro and/or scripting language.

Permitted Uses:
- Licensee may use the number of copies of the Software, Data, and Related Materials for which license fees have been paid on the computer system(s) and/or specific computer network(s) for Licensee's own internal use. Licensee may use the Software, Data, and Related Materials as a map/data server engine in an Internet and/or Intranet distributed computing network or environment provided the appropriate, additional license fees are paid. If the Software, Data, and Related Materials contain dual media (i.e., both 3.5-inch diskettes and CD–ROM), then Licensee may only use one (1) set of the dual media provided. Licensee may not use the other media on another computer system(s) and/or specific computer network(s), or loan, rent, lease, or transfer the other media to another user.
- Licensee may install the number of copies of the Software, Data, and Related Materials for which license or update fees have been paid onto the permanent storage device(s) on the computer system(s) and/or specific computer network(s).
- Licensee may make routine computer backups but only one (1) copy of the Software, Data, and Related Materials for archival purposes during the term of this Agreement unless the right to make additional copies is granted to Licensee in writing by ESRI.
- Licensee may use, copy, alter, modify, merge, reproduce, and/or create derivative works of the online documentation for Licensee's own internal use. The portions of the online documentation merged with other software, hard copy, and/or digital materials shall continue to be subject to the terms and conditions of this Agreement and shall provide the following copyright attribution notice acknowledging ESRI's proprietary rights in the online documentation: "Portions of this document include intellectual property of ESRI and are used herein by permission. Copyright © 200_ Environmental Systems Research Institute, Inc. All Rights Reserved."
- Licensee may use the Data that are provided under license from ESRI and its third party licensor(s) as described in the Distribution Rights section of the online Data Help files.

Uses Not Permitted:
- Licensee shall not sell, rent, lease, sublicense, lend, assign, time-share, or transfer, in whole or in part, or provide unlicensed third parties access to prior or present versions of the Software, Data, and Related Materials, any updates, or Licensee's rights under this Agreement.
- Licensee shall not reverse engineer, decompile, or disassemble the Software, or make any attempt to unlock or bypass the software keycode and/or hardware key used, as applicable, subject to local law.
- Licensee shall not make additional copies of the Software, Data, and/or Related Materials beyond that described in the Permitted Uses section above.
- Licensee shall not remove or obscure any ESRI copyright or trademark notices.
- Licensee shall not use this software for more than one hundred twenty (120) days from the date that the software was installed. At the end of this period, users must remove the time limited software from their computers or purchase fully licensed software. Students and instructors in the United States may purchase fully licensed individual copies of the software from ESRI telesales at 1-800-GIS-XPRT.

Term: The license granted by this Agreement shall commence upon Licensee's receipt of the Software, Data, and Related Materials and shall continue until such time that (1) Licensee elects to discontinue use of the Software, Data, and Related Materials and terminates this Agreement or (2) ESRI terminates for Licensee's material breach of this Agreement. Upon termination of this Agreement in either instance, Licensee shall return to ESRI the Software, Data, Related Materials, and any whole or partial copies, codes, modifications, and merged portions in any form. The parties hereby agree that all provisions, which operate to protect the rights of ESRI, shall remain in force should breach occur.

Limited Warranty: ESRI warrants that the media upon which the Software, Data, and Related Materials are provided will be free from defects in materials and workmanship under normal use and service for a period of sixty (60) days from the date of receipt. The Data herein have been obtained from sources believed to be reliable, but its accuracy and completeness, and the opinions based thereon, are not guaranteed. Every effort has been made to provide accurate Data in this package. The Licensee acknowledges that the Data may contain some nonconformities, defects, errors, and/or omissions. ESRI and third party licensor(s) do not warrant that the Data will meet Licensee's needs or expectations, that the use of the Data will be uninterrupted, or that all nonconformities can or will be corrected. ESRI and the respective third party licensor(s) are not inviting reliance on these Data, and Licensee should always verify actual map data and information. The Data contained in this package are subject to change without notice.

EXCEPT FOR THE ABOVE EXPRESS LIMITED WARRANTIES, THE SOFTWARE, DATA, AND RELATED MATERIALS CONTAINED THEREIN ARE PROVIDED "AS IS," WITHOUT WARRANTY OF ANY KIND, EITHER EXPRESS OR IMPLIED, INCLUDING, BUT NOT LIMITED TO, THE IMPLIED WARRANTIES OF MERCHANTABILITY AND FITNESS FOR A PARTICULAR PURPOSE.

Exclusive Remedy and Limitation of Liability: During the warranty period, ESRI's entire liability and Licensee's exclusive remedy shall be the return of the license fee paid for the Software, Data, and Related Materials in accordance with the ESRI Customer Assurance Program for the Software, Data, and Related Materials that do not meet ESRI's Limited Warranty and that are returned to ESRI or its dealers with a copy of Licensee's proof of payment.

ESRI shall not be liable for indirect, special, incidental, or consequential damages related to Licensee's use of the Software, Data, and Related Materials, even if ESRI is advised of the possibility of such damage.

Waivers: No failure or delay by ESRI in enforcing any right or remedy under this Agreement shall be construed as a waiver of any future or other exercise of such right or remedy by ESRI.

Order of Precedence: Any conflict and/or inconsistency between the terms of this Agreement and any FAR, DFAR, purchase order, or other terms shall be resolved in favor of the terms expressed in this Agreement, subject to the U.S. Government's minimum rights unless agreed otherwise.

Export Regulations: Licensee acknowledges that this Agreement and the performance thereof are subject to compliance with any and all applicable United States laws, regulations, or orders relating to the export of computer software or know-how relating thereto. ESRI Software, Data, and Related Materials have been determined to be Technical Data under United States export laws. Licensee agrees to comply with all laws, regulations, and orders of the United States in regard to any export of such Technical Data. Licensee agrees not to disclose or reexport any Technical Data received under this Agreement in or to any countries for which the United States Government requires an export license or other supporting documentation at the time of export or transfer, unless Licensee has obtained prior written authorization from ESRI and the U.S. Office of Export Control. The countries restricted at the time of this Agreement are Cuba, Iran, Iraq, Libya, North Korea, Serbia, and Sudan.

U.S. Government Restricted/Limited Rights: Any software, documentation, and/or data delivered hereunder is subject to the terms of the License Agreement. In no event shall the Government acquire greater than RESTRICTED/LIMITED RIGHTS. At a minimum, use, duplication, or disclosure by the Government is subject to restrictions as set forth in FAR §52.227-14 Alternates I, II, and III (JUN 1987); FAR §52.227-19 (JUN 1987) and/or FAR §12.211/12.212 (Commercial Technical Data/Computer Software); and DFARS §252.227-7015 (NOV 1995) (Technical Data) and/or DFARS §227.7202 (Computer Software), as applicable. Contractor/Manufacturer is Environmental Systems Research Institute, Inc., 380 New York Street, Redlands, CA 92373-8100, USA.

Governing Law: This Agreement is governed by the laws of the United States of America and the State of California without reference to conflict of laws principles.

Entire Agreement: The parties agree that this constitutes the sole and entire agreement of the parties as to the matter set forth herein and supersedes any previous agreements, understandings, and arrangements between the parties relating hereto and is effective, valid, and binding upon the parties.

ESRI is a trademark of Environmental Systems Research Institute, Inc., registered in the United States and certain other countries; registration is pending in the European Community. SML, AtlasWare, and Avenue are trademarks of Environmental Systems Research Institute, Inc.

Exploring Tropical Cyclones

Introduction
Thinking scientifically iii
Planning to learn iii
GIS made easier iv
Using these materials iv

Getting started
What you need to know v
Minimum system requirements v
Macintosh software installation vi
Using ArcView GIS for Macintosh vii
Windows software installation viii
Using ArcView GIS for Windows ix
Accessing the module data x
ArcView troubleshooting xi
Updates and resources xii

Unit 1 - Recipe for a Cyclone
Activity 1.1   The Great Hurricane of 1900 3
Activity 1.2   Discovering cyclone patterns 11
Activity 1.3   Understanding tropical cyclone physics 15
Activity 1.4   Powering tropical cyclones 19
Activity 1.5   Solving the cyclone puzzle 23

Unit 2 - The Life of a Cyclone
Activity 2.1   Observing tropical cyclones 27
Activity 2.2   Tracking Hurricane Georges 29
Activity 2.3   Classifying tropical cyclones 35
Activity 2.4   Monitoring cyclone growth 39

Unit 3 - Hurricane Hazards

Activity 3.1	Sources of hurricane risk	45
Activity 3.2	The top ten US hurricanes	47
Activity 3.3	Exploring hurricane hazards	51
Activity 3.4	Risk to coastal communities	55

Unit 4 - Hurricanes in the Big Apple

Activity 4.1	Analyzing physical factors	59
Activity 4.2	Managing emergencies	67
Activity 4.3	Addressing demographic factors	69
Activity 4.4	Assessing infrastructure	73

Introduction

Thinking scientifically

Earth scientists make a living by observing and measuring nature. Whether recording the path of a tropical cyclone or mapping the flooding from a major hurricane, a successful Earth scientist relies heavily on his or her ability to recognize patterns. Patterns in space and time are the keys to many of the great discoveries about how the Earth works. The activities in this module will help you develop your ability to recognize and interpret some of nature's fundamental patterns by exploring modern scientific data using a geographic information system (GIS).

Through these exercises you will examine patterns characteristic of tropical cyclones—where and when they form, the sea surface temperatures over which they form, and the directions they move once formed. These patterns will help you to better understand some of the fundamental dynamics that drive Earth's weather, including the storage and transfer of heat energy between the oceans and the atmosphere and the effect of Earth's rotation on air masses (the Coriolis effect).

Most of these patterns are presented in maps, which are one of the Earth scientist's most important tools. The maps will allow you to explore the relationships between natural features and phenomena such as sea surface temperature and a hurricane's wind speed. Overlaying maps of cyclone tracks with maps of features such as roads, cities, and demographics allows you to see the impact of the storms on society. By investigating patterns in a map that shows both topography and areas of predicted hurricane flooding, you can deduce the nature of the relationship between these features. Maps also provide a convenient way to present statistical relationships and maps of different time intervals allow you to look at changes in features over time. You will even be able to create your own maps to explore!

What's a GIS?

A Geographic Information System (GIS) is a tool for organizing, manipulating, analyzing, and visualizing information about the world using a computer.

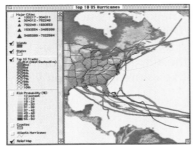

This map shows the tracks of the ten most destructive hurricanes to hit the US.

Demographics = characteristics of human populations such as size, growth, density, distribution, and vital statistics.

A tabular form of the hurricane data shown in the map above. The yellow row is the entry for a single hurricane, Camille.

Planning to learn

Each unit will take you through a well-tested learning process that helps you examine your existing knowledge and build upon it. The first activity will get you thinking about your present knowledge of the major concepts in the unit. It may stimulate questions that you have about the topic. Write these down and, as you learn more, see if you can answer them for yourself. In the second activity you will explore maps and data looking for patterns.

When exploring a particular feature, ask questions like:

- Where do they occur?
- Where don't they occur?
- Why are they there and not elsewhere?
- Are they distributed regularly or irregularly?
- What might cause this pattern?

The third activity provides readings with key information about the major concepts and should help you begin to answer the questions raised above. The readings have been kept brief and provide just the important points you need to continue on or check your prior answers. In the fourth activity, you will apply your new knowledge to solve a particular problem. This will help you build confidence that you understand the material and prepare you for any assessment your instructor may use.

GIS made easier

The purpose of these activities is not to learn how to use a GIS, but to use one as a tool to explore and learn about natural processes and features and how they relate to humans and human activities. For this reason, all of the data have been assembled into ready-to-use projects and complex operations have been eliminated or simplified. While it's good to have basic computer skills, you don't need experience with ArcView GIS software to do these activities. Directions for each new task are provided in the text, so you will learn to use the tool as you explore with it. The activities barely scratch the surface of the data that have been provided, and we encourage you to explore the data on your own and make your own discoveries.

Want to know more?

If you would like more information on how to use ArcView GIS, refer to ***Guide to ArcView GIS***, found in the **Docs** folder on the CD.

Using these materials

Visual cues are used to make the activity directions easier to follow.

- A line preceded by the ▶ symbol is an instruction - something to do on the computer.
- When referring to a tool or button, the name of the tool or button is capitalized and is followed by a picture of that item as it appears on screen—e.g. ...the Identify tool 🛈.
- The ▶ symbol between two boldface words in text indicates a menu choice. Thus, **Theme ▶ Properties** means "pull down the Theme menu and choose Properties."

Theme ▶ Properties *means...*

The most common mistake made when using ArcView GIS software is not selecting or activating the correct theme (map layer) before performing an operation. When things don't seem to be going as they should, this is the first thing to check.

To activate a theme, click on its name in the Table of Contents. Active themes are indicated by a raised border.

The second most common problem is not looking closely enough at the data to see what's going on. ArcView has several tools and buttons for zooming in and out of the map view that work just like tools you've used in other applications—use them!

To zoom in on an area, click and drag with the Zoom In tool 🔍 to outline the area. **To zoom out**, click anywhere on the map window with the Zoom Out tool 🔍.

Getting started

The Exploring Tropical Cyclones CD contains all the software and data you need to complete the module activities on your own Macintosh® or Windows® computer. If you are using this manual as part of a laboratory course, a computer lab with the necessary software and data files may have already been prepared for you in advance.

Additional resources
Visit The SAGUARO Project website for updates, references, and links to related websites:

http://saguaro.geo.arizona.edu

What you need to know

The authors of this book assume that you know how to use a computer with either the Macintosh or Windows operating system installed. We will make no attempt to teach these basic skills:

- turning on the computer and, if necessary, logging in as a user;
- navigating the computer's file system to find folders, applications, and files;
- launching applications and opening files from within those applications; and
- using basic interface elements—opening, closing, moving, and resizing windows, using tools, menus, and dialog boxes, etc.

Installation courtesy
Install the Exploring Tropical Cyclones applications and data only on your own personal computer. If someone else owns the machine, you should seek permission before installing anything.

The data files for this module may be accessed directly from the CD or copied to a user-specified drive and directory during the installation process. Thus, in our modules, you will be instructed to launch the ArcView application, then locate and open a specific file. In a lab setting, your instructor will tell you where to find this file.

Minimum system requirements

Macintosh

- ArcView® GIS 3.0a for Macintosh®
- QuickTime™ and Acrobat® Reader (installers included on CD)
- 133 MHz or faster PowerPC® CPU running Mac OS 8.0 or newer
- 64 MB total RAM (32 MB of available application RAM)
- CD-ROM drive
- 40 MB available hard disk space (+ 100 MB for data, if installed on hard disk)

Mac OS X compatibility
ArcView GIS 3.0a for Macintosh runs in Classic mode under Mac OS X. For best results, restart your computer in OS 9 and install ArcView GIS.

Windows

- ArcView® GIS 3.0a or higher for Windows®
- QuickTime™ and Acrobat® Reader (installers included on CD)
- 133 MHz or faster Pentium™-class CPU running Windows 95 or newer
- 64 MB total RAM (32 MB of available application RAM)
- CD-ROM drive
- 40 MB available hard disk space (+ 100 MB for data, if installed on hard disk)

Macintosh software installation

The instructions below are for installing software on your personal Macintosh computer. If the software has already been installed, such as in a lab setting, skip ahead to the **Using ArcView GIS for Macintosh** section on page vii.

Before you install software

- Insert the **Exploring Tropical Cyclones** CD into your CD-ROM drive and read the **ReadMe** file on the CD for installation updates.
- Disable virus protection software (if installed) and quit any open applications.
- You must have at least 40 MB of free space on your hard drive to install the application. Installing the module data on your hard drive is optional and requires an additional 100 MB of drive space.

Installing ArcView

- Open the **ArcView** folder, double-click the **ArcView Installer** icon, and follow the on-screen instructions.
- When installation has finished, restart your computer and hold down the Command (⌘) and Option keys until the message "Are you sure you want to rebuild the desktop file on the disk..." message appears. Click **OK**. Your computer will rebuild its desktop file and complete the startup process.

Setting ArcView's memory allocation

The default memory allocation for ArcView must be increased to assure trouble-free operation! Follow these steps to increase the application's memory allocation.

- Navigate to where you installed the ArcView GIS application. Look for a folder named **ESRI**—open it and open the folder inside it named **AV_GIS30a**.
- Single-click the **ArcView** application icon to select it.
- Choose **File ▶ Get Info**, choose **Memory** from the **Show** pop-up menu.

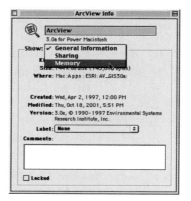

- Enter 32000 for the **Preferred Size** and at least 32000 for the **Minimum Size**. (Note—don't include commas!)

Macintosh CD-ROM Contents

The **Exploring Tropical Cyclones** CD-ROM contains the following folders and files.

Readme.txt
License.txt
Tropical_Cyclones
   cyclones.apr
   sagemdia.txt
   Data (many files)
   Media (many files)
ArcView
   ArcView Installer
Reader
   Acrobat Reader Installer
QuickTime
   QuickTime Installer
Docs
   Guide to ArcView GIS.pdf
   Data Dictionary.pdf

"Can I allocate more memory to ArcView?"

If your computer has more memory available, you can allocate some of the additional memory to ArcView, keeping in mind your other system requirements.

Exploring Tropical Cyclones

Getting started

Where is ArcView GIS?
On a Macintosh, ArcView GIS is installed in a folder named **ESRI** on the drive you specified during installation. You will find the ArcView application in the **AV_GIS30a** folder.

Make an alias!
For convenience, you may wish to make an alias of the ArcView application on your desktop. See your documentation for instructions on how to create an alias.

Monitor resolution
The Exploring Tropical Cyclones activities were designed to be used on a computer with a monitor resolution of at least 800 by 600 pixels and 256 colors. Be sure to set your monitor accordingly, if necessary.

Changing monitor resolution on a Macintosh computer
Select Monitors from the Control Panels under the Apple menu. (Your particular computer may say "Monitors and Sound.") Select the Monitor button and choose the appropriate screen resolution.

Registering ArcView GIS
To register the ArcView GIS application (Macintosh only):
- Double-click the **ArcView** icon to launch the application.
- When prompted, enter your name and company (you may leave this blank), then enter the registration number: **708301184404**.
- Quit ArcView.

Installing QuickTime
The investigations in this module use QuickTime for displaying movies and animations. Most Macintosh computers already have QuickTime installed. If yours does not, you should install it.
- Turn on your computer and insert the **Exploring Tropical Cyclones** CD into the CD-ROM drive.
- Open the **QuickTime** folder on the CD and double-click the **Quicktime Installer** icon. Follow the on-screen instructions to complete the installation.
- Restart your computer.

Installing Acrobat Reader
Acrobat Reader is used to view and print the files in the Docs folder. If Acrobat Reader is already installed on your computer or you do not wish to view or print these documents, you may skip this installation.
- Turn on your computer and insert the **Exploring Tropical Cyclones** CD into the CD-ROM drive.
- Open the **Reader** folder on the CD and double-click the Acrobat Reader Installer icon. Follow the on-screen instructions to complete the installation.
- Restart your computer.

Using ArcView GIS for Macintosh

Launching ArcView and opening project files
- Double-click the ArcView application icon to launch ArcView.
- Choose **File ▶ Open**.
- Navigate to the **Tropical_Cyclones** folder (on your **Exploring Tropical Cyclones** CD or installed on your hard drive), select the ArcView project file **cyclones.apr**, and click **Open**. ArcView project files end in ".apr."

Closing project files
When you have completed an activity or must stop for some reason,
- Choose **File ▶ Quit**.
- When asked if you want to save your changes, click **No**. (Don't worry if you click **Yes**. The files have been locked to prevent accidentally modifying or erasing them.)

Windows software installation

The instructions below are for installing software on your personal Windows-based computer. If the software has already been installed, such as in a lab setting, skip ahead to the **Using ArcView GIS for Windows** section on page ix.

Before you install software

- Disable virus protection software (if installed) and quit any open applications.
- Make sure you have sufficient hard disk space for the installation—40 MB for the application. Installing the module data on your hard disk is optional and requires approximately 100 MB of additional hard disk space.
- You may install this version of ArcView alongside older or newer versions by installing it in a different location on your hard disk.

Installing ArcView GIS

- **NOTE—This is a 120-day license. Do not install until you need to use it, or your license may expire before the semester is over!**
- Insert the **Exploring Tropical Cyclones** CD into your CD-ROM drive.
- Open the ArcView folder, double-click the **AV32AVC.EXE** icon, and follow the on-screen instructions to complete the installation. *(Note - newer versions of Windows may be configured to hide the three character file extension, so this file would appear as simply AV32AVC.)*
- Restart your computer.

Installing QuickTime for Windows

This module uses QuickTime for displaying animations and movies. To install the QuickTime Player application:

- Turn on your computer and insert the **Exploring Tropical Cyclones** CD into the CD-ROM drive.
- Open the **QuickTime** folder on the CD and double-click the QuickTimeInstaller.exe icon. Follow the on-screen instructions to complete the installation.
- Restart your computer.

Installing Acrobat Reader for Windows

Acrobat Reader is used to view and print the files in the Docs folder. If Acrobat Reader is already installed on your computer or you do not wish to view or print these documents, you may skip this installation.

- Turn on your computer and insert the **Exploring Tropical Cyclones** CD into the CD-ROM drive.
- Open the **Reader** folder on the CD and double-click the RP505ENU.EXE icon. Follow the on-screen instructions to complete the installation.
- Restart your computer.

Windows CD-ROM Contents

The Exploring the Dynamic Earth CD-ROM contains the following folders and files.

Readme.txt
License.txt
Tropical_Cyclones
  cyclones.apr
  sagemdia.txt
  Data (many files)
  Media (many files)
ArcView
  AV32AVC.EXE
Reader
  RP505ENU.EXE
QuickTime
  QuickTimeInstaller.exe
Docs
  Guide to ArcView GIS.pdf
  Data Dictionary.pdf

Monitor resolution

The Exploring the Dynamic Earth module was designed to be used with a monitor resolution of at least 800 by 600 pixels and 256 colors.

Changing monitor resolution on a Windows computer

Right-click on the desktop, choose Settings from the popup menu, and click the Settings tab. Set the color palette and desktop area to appropriate values.

For more help, consult your computer's printed or online documentation.

Using ArcView GIS for Windows

Launching ArcView GIS for Windows

- On the Windows desktop, click the **Start** button.
- On the **Start** menu, choose **Programs ▸ ESRI ▸ ArcView GIS Virtual Campus Edition ▸ ArcView GIS Virtual Campus Edition**.
- On the next screen, click the **Try** button. This will begin your 120-day license period. Each time you launch the ArcView GIS application, this screen tells you the number of days remaining in the license period.

Create a shortcut

You may want to create a shortcut to ArcView on your desktop, to make it easier to access the program. To find out how to create a shortcut, refer to your Windows documentation.

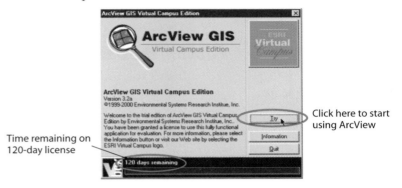

Opening a project file

- Launch ArcView GIS for Windows (see above).
- In the **Welcome to ArcView GIS** dialog box, click the **Open an existing project** button, then click **OK**.
- Use the **Open** dialog box to locate the file you want to open:
  1) Select the drive where you installed the module files.
  2) Select the directory (folder) containing the project files.
  3) Select the file. ArcView project files end in ".apr."
  4) Click **OK**.

Where is ArcView?

The installer places ArcView in the **Program Files** folder of the drive you specified, in a folder named **VCampus**.

- When the project files have opened, you will see the message shown at left. Answer either "yes" or "no"—it makes no difference.

Closing a project file

When you have completed an activity or must stop for some reason,

- Choose **File ▸ Quit**.
- When asked if you want to save your changes, click **No**. (Don't worry if you click **Yes**. The files have been locked to prevent accidentally modifying or erasing them.)

Exploring Tropical Cyclones — Getting started

Accessing the module data

If ArcView and QuickTime software are already installed on your computer, you are ready to use the module. You may access the module data directly from the CD-ROM or copy it to your computer's hard drive for increased convenience and performance.

Reading the data files directly from the CD

- Insert the **Exploring Tropical Cyclones** CD in your CD-ROM drive.
- Launch ArcView, choose **File ▶ Open**, navigate to the **Tropical_Cyclones** folder on the CD, and open the desired ArcView project file.

Copying the data files to your computer

- Insert the **Exploring Tropical Cyclones** CD in your CD-ROM drive.
- Copy the *entire* **Tropical_Cyclones** folder from the CD to your hard disk.
- **Important** - Do not change the name of the **Tropical_Cyclones** folder or any of its contents at any time.
- **Important** - If you are copying the **Tropical_Cyclones** folder inside another folder, give the folder a short name (8 characters or fewer) and do not nest the folder too many levels deep. This will help ensure that ArcView can locate the files correctly.
- **Critical issue for Windows installations only** - There must be *no spaces* in the names of the drive or folders along the path to the **Tropical_Cyclones** folder. If necessary, change spaces in drive and folder names to underscore characters—thus, a **Class Data** folder should be renamed **Class_Data**.
- If the data files get renamed or damaged, delete the entire **Tropical_Cyclones** folder and copy a clean version of the folder from the CD to your hard disk.

Q&A about the 120-day ArcView license on the CD

"My university has an ArcView site license—do I need to use the 120-day version of ArcView on the Exploring the Dynamic Earth CD?"

Only if you want to use your own computer to complete the GIS activities—the 120-day version of ArcView is primarily intended for use on students' personal computers.

"When should I install the 120-day version of ArcView?"

Do not install your ArcView GIS software until you need to use it. The ArcView license allows you to install the ArcView application on one computer for 120 days. At the end of that time, the software will no longer function, and it cannot be reinstalled on that computer.

"Can I reinstall the application if it gets damaged or someone accidentally uninstalls it?"

Yes, but it will only function for 120 days from the original installation date.

"What happens at the end of the 120-day license period?"

Your copy of ArcView GIS will stop working. According to the license agreement, you must uninstall the ArcView GIS application, since it will no longer function and will not work if reinstalled. For pricing and information about student and instructor licenses for ArcView, call ESRI telesales at (800) 447-9778.

"What happens if I retake this course or take a different course using another module from the Exploring series?"

You must install your "new" 120-day licensed version of ArcView on a different computer than your original installation.

ArcView troubleshooting

To activate a theme, click on its name in the Table of Contents. Active themes are indicated by a raised border.

"When I _____ in ArcView, nothing (or the wrong thing) happens."

Most ArcView errors are caused by not having the correct theme activated when performing an operation. By activating a theme, you are telling ArcView which data to operate on. If the wrong data are identified, the operation will most likely fail or produce unexpected results.

"ArcView crashes while launching." (Macintosh and Windows)

- ArcView may not have enough memory to work properly (Macintosh only)—see page vi for directions on **Setting ArcView's memory allocation**.
- ArcView may be damaged. You may uninstall and reinstall ArcView at any time within the 120-day license period using the **Exploring Tropical Cyclones** CD.

"When I open a project file, ArcView keeps telling me it can't find files that I know are installed." (Macintosh and Windows)

The project file contains a path to the location of each data file in the project. If you move files or rename any of the files or folders that were installed, ArcView cannot find the files. If you cannot restore the correct file and folder names and locations, delete the **Tropical_Cyclones** folder and re-install it on the hard disk.

"When I open a project file from within ArcView, it tells me it can't find the project file." (Windows only)

There is probably a space in the name of the drive or folder into which you copied the **Tropical_Cyclones** folder. ArcView for Windows cannot locate a file correctly if it encounters a space in the file's pathname.

"ArcView launches and opens the project file, but it crashes later, after several minutes of use ." (Mac)

The memory allocation for ArcView is probably too low. See page vi for directions on **Setting ArcView's memory allocation**.

"When I use ArcView's Media Viewer to view movies or animations, I get a message inviting me to upgrade to QuickTime Pro. When I click the "later" button, nothing happens." (Windows only)

This usually occurs only in a lab or other shared environment, when the QuickTime preferences file is not "writable" to all users. The preferences file is called **QuickTime.qtp**; its exact location depends on the version of Windows you are using. Consult your system administrator about making the necessary changes.

Updates and resources

For corrections, updates, and learning tips for this module, visit the SAGUARO Project website at:

http://saguaro.geo.arizona.edu

Online help

Two Help menus on a Mac?
On the Macintosh, ArcView adds a second Help menu of its own to the left of the standard Macintosh Help menu. Use this menu to access ArcView's built-in help system.

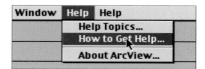

This module provides all of the directions you need to complete the activities using ArcView GIS. If you wish to explore ArcView on your own, or learn more about ArcView's capabilities, you may wish to consult ArcView's **Help** menu. Choose **Help ▶ How to Get Help...** to learn more about using ArcView's built-in help system.

Printing the Educator's Guide to ArcView GIS

The **Docs** folder contains a handy quick reference manual of the most common tools and techniques used in ArcView, called the Educator's Guide to ArcView GIS. You may view this document on screen or print all or part of it for later reference using the included Acrobat Reader software.

- Turn on your computer and insert the **Exploring Tropical Cyclones** CD into the CD-ROM drive.
- Open the **Docs** folder on the CD and double-click the **Guide to ArcView GIS.pdf** icon.
- Use the on-screen controls to scroll through the document, and choose **File ▶ Print** and either print all pages or select a range of pages to print.

The Data Dictionary file

The **Data Dictionary.pdf** file found in the **Docs** folder provides information about the data used in this module, including a description of where the data came from and how they were processed. You do not need this information to complete the activities in the module, but you may find it useful if you are using the data for independent research or just to satisfy your curiosity. Many of the data providers listed are good sources for additional data and information.

Unit 1
Recipe for a Cyclone

In this unit, you will learn...

- *what caused the worst natural disaster ever on American soil,*
- *where and when tropical cyclones form,*
- *the source of energy that powers tropical cyclones,*
- *what causes tropical cyclones to spin, and*
- *what ingredients it takes to create a tropical cyclone.*

Exploring Tropical Cyclones | **Unit 1 - Recipe for a Cyclone**

Activity 1.1

The Great Hurricane of 1900

On September 8, 1900, the greatest natural disaster to strike the United States occurred at Galveston, Texas. In the early evening hours of September 8, a hurricane made landfall, bringing with it a 5-meter (17-foot) storm surge that inundated most of Galveston Island and the city of Galveston. By the next day, much of the city was destroyed, at least 8,000 people were killed, and many thousands more were made homeless.

The account beginning on page 5 is an eyewitness report of the storm and its aftermath by Isaac M. Cline, the senior Weather Bureau employee at Galveston in 1900.

1900 Galveston Hurricane vital statistics

- **Maximum wind speed:** 120 mph (194 km/hr)[a]
- **Storm surge:** 15 to 20 ft (5-6 meters)
- **Galveston's highest point:** 8.7 feet above sea level (2 meters)
- **Local tide at time of storm surge:** high
- **Lowest observed air pressure at the Galveston weather office:** 28.53 inches Hg (966 hPa)[b]
- **Lowest observed air pressure at sea:** 27.49 inches Hg (931 hPa)[c]
- **Estimated Intensity:** Category 4 hurricane
- **Population of Galveston:** 37,789 (1900 Census)
- **Fatalities:** Estimated at 6,000 - 8,000 in Galveston plus 2,000 in surrounding area. Some place the figure as high as 12,000.
- **Number of homes destroyed:** over 3,600 homes in Galveston (estimate)
- **Total damage :** $30 million (estimate - equivalent to $620 million in 2000 US dollars)

Hurricane behavior and hazards

Hurricanes, also known as tropical cyclones, unleash incredible amounts of energy over wide areas, and are capable of tremendous destruction. After reading the story of the 1900 Galveston Hurricane, list and describe all of the hurricane-related hazards mentioned in the story. Feel free to add other hazards from your previous knowledge or experience with tropical cyclones. Be prepared to discuss your list with your classmates.

1.
2.
3.
4.
5.
6.
7.
8.

[a] Estimated speed - the official anemometer blew away after recording a sustained wind speed of 84 mph (135 km/hr) and gusts of 102 mph (165 km/hr).

[b] **Inches Hg** stands for inches of mercury, a standard unit of atmospheric pressure. Hg is the chemical symbol for mercury, the silvery liquid metal once used in barometers.

[c] The abbreviation **hPa** stands for hectopascals, a metric unit of pressure.

anemometer (NOAA/NWS)

Questions

1. Which of the hazards you listed was responsible for the greatest damage and loss of life?

2. Why didn't the people of Galveston evacuate the city before the hurricane struck?

3. What, if anything, do you think the city of Galveston could have done to prevent the high death toll and property damage caused by the 1900 hurricane?

4. If this hurricane struck Galveston Island today, would it have the same destructive effect? Explain.

5. With today's technology, do you think a tropical cyclone could cause a disaster of this magnitude anywhere around the world? Explain.

Eyewitness account by Dr. Isaac M. Cline
Special Report on the Galveston Hurricane

Isaac M. Cline, Local Forecast Official and Section Director
SEPTEMBER 8, 1900

The hurricane which visited Galveston Island on Saturday, September 8, 1900, was no doubt one of the most important meteorological events in the world's history. The ruin which it wrought beggars description, and conservative estimates place the loss of life at the appalling figure, 6,000.

A brief description of Galveston Island will not be out of place as introductory to the details of this disaster. It is a sand island about thirty miles in length and one and one-half to three miles in width. The course of the island is southwest to northeast, parallel with the southeast coast of the State. The City of Galveston is located on the east end of the island. To the northeast of Galveston is Bolivar Peninsula, a sand spit about twenty miles in length and varying in width from one-fourth of a mile to about three miles. Inside of Galveston Island and Bolivar Peninsula is Galveston bay, a shallow body of water with an area of nearly five hundred square miles. The length of the bay along shore is about fifty miles and its greatest distance from the Gulf coast is about twenty-five miles. The greater portion of the bay lies due north of Galveston. That portion of the bay which separates the island west of Galveston from the mainland is very narrow, being only about two miles in width in places, and discharges into the Gulf of Mexico through San Louis Pass. The main bay discharges into the Gulf between the jetties; the south one being built out from the northeast end of Galveston Island and the north one from the most southerly point of Bolivar Peninsula. The channel between the jetties is twenty-seven to thirty feet in depth at different stages of the tide. There are channels in the harbor with a depth of thirty to thirty-five feet, and there is an area of nearly two thousand acres with an anchorage depth of eighteen feet or more. The mainland for several miles back of the bay is very low, in fact much of it is lower than Galveston Island, and it is so frequently overflowed by high tide that large areas present a marshy appearance. These are in brief the physical conditions of the territory devastated by the hurricane.

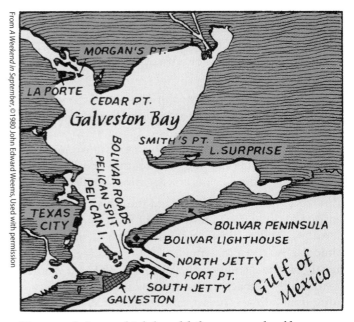

The usual signs which herald the approach of hurricanes were not present in this case. The brick-dust sky was not in evidence to the smallest degree. This feature, which has been distinctly observed in other storms that have occurred in this section, was carefully watched for, both on the evening of the 7th and the morning of the 8th. There were cirrus clouds moving from the southeast during the forenoon of the 7th, but by noon only altostratus from the northeast were observed. About the middle of the afternoon the clouds were divided between cirrus, altostratus, and cumulus, moving from the northeast. A heavy swell from the southeast made its appearance in the Gulf of Mexico during the afternoon of the 7th. The swell continued during the night without diminishing, and the tide rose to an unusual height when it is considered that the wind was from the north and northwest. About 5 AM of the 8th Mr. J. L. Cline, Observer, called me and stated that the tide was well up in the low parts of the city, and that we might be able to telegraph important information to Washington. He having been on duty until nearly midnight, was told to retire and I would look into conditions. I drove to the Gulf, where I timed the swells, and then proceeded to the office and found that the barometer was only one-tenth of an inch lower than it was at the 8 PM observation of the 7th. I then returned to the Gulf, made more detailed observations of the tide and swells, and filed the following telegram addressed to the Central Office in Washington:

> "Unusually heavy swells from the southeast, intervals of one to five minutes, overflowing low places south portion of city three to four blocks from beach. Such high water with opposing winds never observed previously.
>
> Broken stratus and stratocumulus clouds predominated during the early forenoon of the 8th, with the blue sky visible here and there. Showery weather commenced at 8:45 AM, but dense clouds and heavy rain were not in evidence until about noon, after which dense clouds with rain prevailed."

The wind during the forenoon of the 8th was generally north, but oscillated, at intervals of from five to ten minutes, between northwest and northeast, and continued so up to 1 PM. After 1 PM, the wind was mostly northeast, although as late as 6:30 PM it would occasionally go back to the northwest for one or two minutes at a time. The prevailing wind was from the northeast until 8:30 PM, when it shifted to the east, continuing from this direction until about 10 PM. After 10 PM the wind was from the southeast, and after about 11 PM the prevailing direction was from the south or southwest. The directions after 11 PM are from personal observations. A storm velocity was not attained until about 1 PM after which the wind increased steadily and reached a hurricane velocity about 5 PM. The greatest velocity for five minutes was 84 miles per hour at 6:15 PM. With two minutes at the rate of 100 miles per hour. The anemometer blew away at this time, and it is estimated that prior to 8 PM the wind attained a velocity of at least 120 miles per hour. For a short time, about 8 PM, just before the wind shifted to the east, there was a distinct lull, but when it came out from the east and southeast it appeared to come with greater fury than before. After shifting to the south at about 11 PM the wind steadily diminished in velocity, and at 8 AM on the morning of the 9th was blowing at the rate of 20 miles per hour from the south.

The barometer commenced falling on the afternoon of the 6th and continued falling steadily but slowly up to noon of the 8th, when it read 29.42 inches. The barometer fell rapidly from noon until 8:30 PM of the 8th, when it registered 28.48 inches, a fall of pressure of about one inch in eight and one-half hours. After 8:30 PM the barometer rose at the same rapid rate that had characterized the fall.

On account of the rapid fall in pressure, Mr. John D. Blagden, observer, took readings of the mercurial barometer as a check on the barograph, and readings are as follows:

Time	Reading	Time	Reading
5:00 PM	29.05	6:40 PM	28.75
5:11 PM	29.00	6:48 PM	28.70
5:30 PM	28.95	7:15 PM	28.69
5:50 PM	28.90	7:40 PM	28.62
6:06 PM	28.86	8:00 PM	28.55
6:20 PM	28.82	8:10 PM	28.53

These readings confirm the low pressure shown by barograph and indicate the great intensity of the hurricane.

Mr. Blagden looked after the instruments during the hurricane in a heroic and commendable manner. He kept the wires of the self-registering apparatus intact as long as it was possible for him to reach the roof. The rain gauge blew away about 6 PM and the thermometer shelter soon followed. All the instruments in the thermometer shelter were broken, except the thermograph which was found damaged, but has been put in working order.

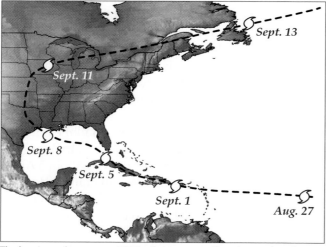

The hurricane, born some 4,000 miles away, was first observed on August 30, east of Puerto Rico. Galvestonians became aware of the storm on September 4, but seriously underestimated its strength. The city was considered safe from such disasters, having survived major storms in 1875 and 1886 with minimal damage and loss of life.

Storm warnings were timely and received a wide distribution not only in Galveston but throughout the coast region. Warning messages were received from the Central Office at Washington on September 4, 5, 6, 7, and 8. The high tide on the morning of the 8th, with storm warning flying, made it necessary to keep one man constantly at the telephone giving out information. Hundreds of people who could not reach us by telephone came to the Weather Bureau office seeking advice. I went down on Strand street and advised some wholesale commission merchants who had perishable goods on their floors to place them 3 feet above the floor. One gentleman has informed me that he carried out my instructions, but the wind blew his goods down. The public was warned, over the telephone and verbally, that the wind would go by the east to the south and that the worst was yet to come. People were advised to seek secure places for the night. As a result thousands of people who lived near the beach or in small houses moved their families into the center of the city and were thus saved. Those who lived in large strong buildings, a few blocks from the beach, one of whom was the writer of this report, thought that they could weather the wind and tide. Soon after 3 PM conditions became so threatening that it was deemed essential that a special report be sent at once to Washington. Mr. J. L. Cline, Observer, took the instrumental readings while I drove first to the bay and then to the Gulf, and finding that half the streets of the city were under water added the following to the special observation at 3:30 PM: "Gulf rising, water covers streets of about half of city." Having been on duty since 5 AM, after giving this message to the observer, I went home to lunch. Mr. J. L. Cline went to the telegraph offices through water from two to four feet deep, and found that the telegraph wires had all gone down; he then returned to the office, and by inquiry learned that the long distance telephone had one wire still working to Houston, over which he gave the message to the Western Union telegraph office at Houston to be forwarded to the Central Office at Washington.

After the disaster, survivors returned to salvage what they could from the debris, but little remained of their former homes and businesses.

I reached home and found the water around my residence waist deep. I at once went to work assisting people, who were not securely located, into my residence, until forty or fifty persons were housed therein. About 6:30 PM Mr. J. L. Cline, who had left Mr. Blagden at the office to look after the instruments, reached my residence, where he found the water neck deep. He informed me that the barometer had fallen below 29.00 inches; that no further messages could be gotten off on account of all wires being down, and that he had advised everyone he could see to go to the center of the city; also, that he thought we had better make an attempt in that direction. At this time, however, the roofs of houses and timbers were flying through the streets as though they were paper, and it appeared suicidal to attempt a journey through the flying timbers. Many people were killed by flying timbers about this time while endeavoring to escape to town.

The water rose at a steady rate from 3 PM until about 7:30 PM, when there was a sudden rise of about four feet in as many seconds. I was standing at my front door, which was partly open, watching the water, which was flowing with great rapidity from east to west. The water at this time was about eight inches deep in my residence, and the sudden rise of 4 feet brought it above my waist before I could change my position. The water had now reached a stage 10 feet above the ground at Rosenberg Avenue (Twenty-fifth street) and Q street,

Many people went to upper floors and climbed onto roofs to escape the rising water. Many wood-frame buildings were knocked from their foundations and disintegrated to become part of the sea of floating debris.

where my residence stood. The ground was 5.2 feet elevation, which made the tide 15.2 feet. The tide rose the next hour, between 7:30 and 8:30 PM, nearly five feet additional, making a total tide in that locality of about twenty feet. These observations were carefully taken and represent to within a few tenths of a foot the true conditions. Other personal observations in my vicinity confirm these estimates. The tide, however, on the bay or north side of the city did not obtain a height of more than 15 feet. It is possible that there was 5 feet of backwater on the Gulf side as a result of debris accumulating four to six blocks inland. The debris is piled eight to fifteen feet in height.

By 8 PM a number of houses had drifted up and lodged to the east and southeast of my residence, and these with the force of the waves acted as a battering ram against which it was impossible for any building to stand for any length of time, and at 8:30 PM my residence went down with about fifty persons who had sought it for safety, and all but eighteen were hurled into eternity. Among the lost was my wife, who never rose above the water after the wreck of the building. I was nearly drowned and became unconscious, but recovered though being crushed by timbers and found myself clinging to my youngest child, who had gone down with myself and wife. Mr. J. L. Cline joined me five minutes later with my other two children, and with them and a woman and child we picked up from the raging waters, we drifted for three hours, landing 300 yards from where we started. There were two hours that we did not see a house nor any person, and from the swell we inferred that we were drifting to sea, which, in view of the northeast wind then blowing, was more than probable. During the last hour that we were drifting, which was with southeast and south winds, the wreckage on which we were floating knocked several residences to pieces. When we landed about 11:30 PM, by climbing over floating debris to a residence on Twenty-eighth street and Avenue P, the water had fallen about 4 feet. It continued falling, and on the following morning the Gulf was nearly normal. While we were drifting we had to protect ourselves from the flying timbers by holding planks between us and the wind, and with this protection we were frequently knocked great distances. Many persons were killed on top of the drifting debris by flying timbers after they had escaped from their wrecked homes. In order to keep on the top of the floating masses of wrecked buildings one had to be constantly on the lookout and continually climbing from drift to drift. Hundreds of people had similar experiences.

Sunday, September 9, 1900, revealed one of the most horrible sights that ever a civilized people looked upon. About three thousand homes, nearly half the residence portion of Galveston, had been completely swept out of existence, and probably more than six thousand persons had passed from life to death during that dreadful night. The correct number of those who perished will probably never be known, for many entire families are missing. Where 20,000 people lived on the 8th not a house remained on the 9th, and who occupied the houses may, in many instances, never be known. On account of the pleasant Gulf breezes many strangers were residing temporarily near the beach, and the number of these that were lost can not yet be estimated. I enclose a chart, fig. 2, which shows, by shading, the area of total destruction. Two charts of this area have been drawn independently; one by Mr. A. G. Youens, inspector for the local board of underwriters, and the other by myself and Mr. J. L. Cline. The two charts agree in nearly all particulars, and it is believed that the chart enclosed represents the true conditions as nearly as it is possible to show them. That portion of the city west of Forty-fifth street was sparsely settled, but there were several splendid residences in the southern part of it. Many truck farmers and dairy men resided on the west end of the island, and it is estimated that half of these were lost, as but very few residences remain standing down the island. For two blocks, inside the shaded area, the damage amounts to at least fifty per cent of the property. There is not a house in Galveston that escaped injury, and there are houses totally wrecked in all parts of the city. All goods and supplies not over eight feet above floor

The dead were carried by wagons to be loaded onto barges for burial at sea. Many bodies later washed ashore, requiring them to be buried again.

were badly injured, and much was totally lost. The damage to buildings, personal, and other property in Galveston County is estimated at above thirty million dollars. The insurance inspector for Galveston states that there were 2,636 residences located prior to the hurricane in the area of total destruction, and he estimates 1,000 houses totally destroyed in other portions of the city, making a total of 3,636 houses totally destroyed. The value of these buildings alone is estimated at $5,500,000.

Many survivors took refuge in a handful of large stone buildings such as churches and hospitals. Here, survivors inspect the devastation.

The grain elevators which were full of grain suffered the smallest damage. Ships have resumed loading and work is being rushed day and night. The railroad bridges across the bay were washed away, but one of these has been repaired and direct rail communication with the outside world was established within eleven days after the disaster. Repairs and extensions of wharves are now being pushed forward with great rapidity. Notwithstanding the fact that the streets are not yet clean and dead bodies are being discovered daily among the drifted debris, the people appear to have confidence in the place and are determined to rebuild and reestablish themselves here. Galveston being one of the richest cities of its size in the United States, there is no question but that business will soon regain its normal condition and the city will grow and prosper as she did before the disaster. Cotton is now coming in by rail from different parts of the State and by barge from Houston. The wheels of commerce are already moving in a manner which gives assurance for the future. Improvements will be made stronger and more judiciously; for the past twenty-five years they have been made with the hurricane of 1875 in mind, but no one ever dreamed that the water would reach the height observed in the present case. The railroad bridges are to be built ten feet higher than they were before. The engineer of the Southern Pacific Company has informed me that they will construct their wharves so that they will withstand even such a hurricane as the one we have just experienced.

I believe that a sea wall, which would have broken the swells, would have saved much loss of both life and property. I base this view upon observations which I have made in the extreme northeastern portion of the city, which is practically protected by the south jetty; this part of the city did not suffer more than half the damage that other similarly located districts, without protection, sustained.

From the officers of the U. S. Engineer tug Anna, I learn that the wind at the mouth of the Brazos River went from north to southwest by way of west. This shows that the center of the hurricane was near Galveston, probably not more than 30 miles to the westward. The following towns have suffered great damage, both in the loss of life and property: Texas City, Dickinson, Lamarque, Hitchcock, Arcadia, Alvin, Manvel, Brazoria, Columbia, and Wharton. Other towns further inland have suffered, but not so seriously. The exact damage at these places can not be ascertained.

A list of those lost in Galveston, whose names have been ascertained up to the present time, contains 3,536 names. *[This was later revised to as many as 8,000.]*

UNITED STATES WEATHER BUREAU OFFICE, GALVESTON, TEX., September 23, 1900.

Discovering cyclone patterns

Activity 1.2

Things to know...

The ▶ symbol between two boldface words in text indicates a menu choice. Thus, **File ▶ Open...** means "pull down the File menu and choose Open... from the menu."

How to zoom in and out. Use the Zoom In and Zoom Out tools. To see the entire map again, choose View > Full Extent.

To turn a theme on or off, click its checkbox in the Table of Contents.

To activate a theme, click on its name in the Table of Contents. Active themes are indicated by a raised border.

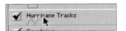

Hurricane Andrew movie

The Hurricane Andrew movie is an animated sequence of GOES weather satellite images, showing the motion and development of the storm.

Reading coordinates: The cursor's longitude and latitude are displayed on the right end of the tool bar.

Tropical cyclones are among the most powerful and destructive natural phenomena on Earth. In one day, a tropical cyclone releases enough energy to supply the electrical needs of the United States for six months! Called hurricanes, cyclones, or typhoons depending on where they occur, these massive storms kill thousands of people and cause billions of dollars in damage each year.

A global view

In this activity, you will investigate where and when tropical cyclones form, to help you understand the conditions that create and maintain these huge storms.

▶ Launch the ArcView GIS application, then locate and open the **cyclones.apr** project file.
▶ Open the **Global Patterns** view.

The green symbols show the starting point of each of the 4332 tropical cyclones recorded between 1950 and 2000. Keep in mind that each dot represents the beginning of a storm that affected a broad area. To illustrate the size of these storms, you will look at an animated series of weather satellite images of North America from 1992.

▶ Click the Media Viewer button and choose the **Hurricane Andrew Movie** from the list. After the QuickTime Player application loads the file, click the play button to view the movie.

The movie covers a six-day period from August 22–27. At the beginning of the movie, you should see Hurricane Lester making landfall on the west coast of Mexico. A day or two later, Hurricane Andrew crosses Florida before slamming into Louisiana.

1. In which direction do Andrew and Lester spin—clockwise or counterclockwise?

2. What happens to both storms when they cross over land?

▶ Close the QuickTime Player application when you are finished viewing the movie.
▶ Now, look at the distribution of hurricanes on the map. Using the latitude lines on the map, estimate the northern and southern boundaries of the region where tropical cyclones form.

3. Most tropical cyclones form between about _____ °N latitude and _____ °S latitude.

Exploring Tropical Cyclones Unit 1 - Recipe for a Cyclone

Zooming in quickly
The quickest way to zoom in on an area is to use the Zoom In tool to click and drag a rectangle around the region you want to fill the view.

Using the Measure tool
To measure the distance between two points, click the Measure tool's crosshair cursor on one point, move the crosshair to the other point, and read the distance in the status bar. Double-click to stop measuring.

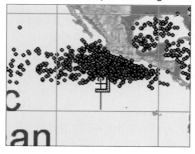

Look closely at the area near the equator. There appears to be a narrow, "cyclone-free zone" centered on the equator.

▶ Using the Zoom In tool, zoom in on this cyclone-free zone, and use the Pan tool to drag the map along the equator.

▶ Move your cursor along the boundaries of the zone and note the latitude in the coordinate display at the right end of the tool bar.

▶ Use the Measure tool to measure the distance from the equator to the edge of the cyclone-free zone, as shown at left. The distance (Length) is given in the status bar.

read distance here (your distance will vary)
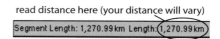

4. The cyclone-free zone extends approximately _____ degrees or _____ kilometers from the equator.

For now, it's enough to know that tropical cyclones don't form in this band. Later, you'll learn *why* they don't form there.

▶ Click the Zoom to Full Extent button to display the entire map again.

5. Symmetrical patterns are very common in nature. In what areas of the world oceans, besides those very near the equator and the poles, would symmetry lead you to expect tropical cyclones to form, yet they don't? Identify these areas on the map below.

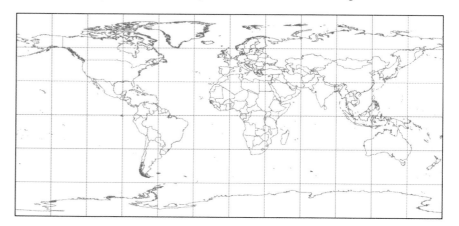

6. Explain why you expect tropical cyclones to form in these areas.

12   Discovering cyclone patterns

Exploring Tropical Cyclones — Unit 1 - Recipe for a Cyclone

Tropical cyclone basins

To turn a theme on or off, click its checkbox in the Table of Contents.

To activate a theme, click on its name in the Table of Contents.

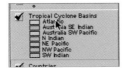

Within the range of latitudes you observed, the tropical cyclones aren't distributed evenly around the globe. Instead, they occur in large clusters. These clusters help define tropical cyclone basins.

▸ Turn off the **Tropical Cyclones** theme.
▸ Turn on and activate the **Tropical Cyclone Basins** theme.

This theme shows seven major regions, called basins, where tropical cyclones form. Each basin is identified by the ocean in which it occurs and the land areas affected by its storms. A storm's name also depends on the basin in which it forms.

▸ To learn more about these basins and answer the questions below:
  • Click on each basin using the Identify tool.
  • Close each **Identify Results** window when you are done with it.

7. In the table, record the hemisphere, direction of rotation, average number of cyclones per year, and storm name for each basin.

Tropical Cyclone Basin Name	Hemisphere (N / S)	Storm rotation (CW / CCW)	Average number of tropical cyclones per pear	Name of storm (from [Type] field)
Atlantic				
Australia SE Indian				
Australia SW Pacific				
N Indian				
NE Pacific				
NW Pacific				
SW Indian				

Using the information from your table, answer the following questions.

8. What are all tropical cyclones in the southern hemisphere called?

9. What would you call a tropical cyclone that strikes Japan—a cyclone, a typhoon, or a hurricane?

10. In which direction do storms in each hemisphere rotate?
    Northern = _____
    Southern = _____

What's in a name?

Cyclone comes from the Greek word *kuklos*, meaning circle. Indeed, tropical cyclones not only spin, but they usually move along curved paths.

Hurricane comes from *Hurakán*, the Mayan god of the skies and lightning. He created the world by endlessly repeating the word "earth" until the solid ground rose from the seas. When he became angry with the first human beings, Hurakán unleashed the rain and wind that destroyed them. Hurakán translates literally as "one-legged."

Typhoon comes from the Japanese word *taifuu*. The characters translate literally as "pedestal wind," a very vivid description!

台 tai = *pedestal*
風 fuu = *wind*

Discovering cyclone patterns 13

Exploring Tropical Cyclones
Unit 1 - Recipe for a Cyclone

When do tropical cyclones occur?

For people living on the Atlantic and Gulf Coasts, a typical year has *five*, not four seasons. The fifth season is hurricane season. With it comes the fear that it might just be the year of "THE BIG ONE." Is tropical cyclone season the same everywhere on Earth? To find out, you will change the legend to show the time of year during which each storm occurred.

▶ Turn on and activate the **Tropical Cyclones** theme.
▶ To change the **Tropical Cyclones** legend:
- Double-click the **Tropical Cyclones** theme title in the legend to open the Legend Editor.
- Click the **Load** button, open the **Data** folder, then open the **seasons.avl** file.
- In the next dialog box, choose **Dates** from the **Field** popup menu and click **OK**.
- Click the **Apply** button, then close the Legend Editor window.

Each tropical cyclone should be colored according to the time of year when it formed. (If not, repeat the process or ask your instructor for help.)

▶ Use the Zoom In tool to magnify each of the tropical cyclone basins. For each basin, note the predominant color of the tropical cyclones and find the corresponding dates in the legend.

11. For each tropical cyclone basin, record the hemisphere and dates of greatest tropical cyclone activity in the table below. Use the information at left to convert each range of dates to a season.

To edit a theme's legend, double-click the name of the theme in the Table of Contents.

Earth's Seasons

Earth's seasons are
- caused by the tilt of Earth's axis,
- defined according to the elevation of the sun in the sky, and
- opposite in the northern and southern hemispheres.

Dates	Hemisphere	
	Northern	Southern
Dec 21 - Mar 19	Winter	Summer
Mar 20 - Jun 20	Spring	Fall
Jun 21 - Sep 21	Summer	Winter
Sep 22 - Dec 20	Fall	Spring

Tropical Cyclone Basin Name	Hemisphere (N / S)	Range of dates of greatest tropical cyclone activity	Season (see table at left)
Atlantic			
Australia SE Indian			
Australia SW Pacific			
N Indian			
NE Pacific			
NW Pacific			
SW Indian			

12. According to the table, regardless of the hemisphere, during which season of the year do most tropical cyclones occur?

14 Discovering cyclone patterns

Activity 1.3
Understanding tropical cyclone physics

Solar-powered storms

In the previous activity, you discovered that most tropical cyclones form during the summer and early fall. This is because tropical cyclones are powered by solar energy, and summer is when Earth receives the most energy from the sun. Summer doesn't occur at the same time of year everywhere and, as you've seen, neither do tropical cyclones. In the northern hemisphere, tropical cyclone season is from June through September. In the southern hemisphere it runs from December through March.

The reason for seasons

The seasons are opposite in each hemisphere for the same reason that the seasons themselves exist; Earth's axis is tilted 23.5° from "vertical" as it orbits the sun.

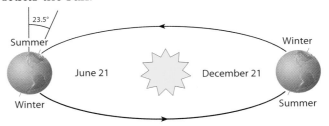

On around June 21st of each year, the north pole tilts most directly toward the sun. This day marks the first day of summer in the northern hemisphere. On the same day, the south pole is pointing most directly away from the sun, marking the first day of winter in that hemisphere. On around December 21, the opposite tilt of the poles marks the first day of northern hemisphere winter and southern hemisphere summer.

The Tropics

Imaginary lines 23.5° north and south of the equator mark where the sun passes overhead on the first day of summer in each hemisphere. These latitudes are called the Tropic of Cancer and the Tropic of Capricorn. The area between these latitudes, called **the tropics**, receives the most direct sunlight throughout the year.

Outside the tropics, the sun never passes directly overhead. Areas north or south of the tropics receive more solar radiation during their summer, when their hemisphere is tilted toward the sun.

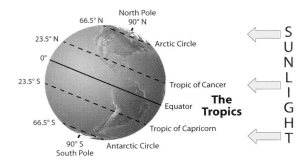

Energy and latitude

In addition to the seasons, the tropics are warm because of the shape of the Earth. As you move toward the poles, the sun's rays strike the ground at lower angles, spreading the same amount of energy over a greater area, as shown below.

This explains the temperature differences at different latitudes. The same amount of sunlight that heats up one square meter at the equator is spread over 1.4 square meters at 45° latitude, 2 square meters at 60° latitude, and over 11 square meters near the poles at 85°. As you move closer to the poles, each square meter receives less solar energy.

1. Where are the Tropics, and why do cyclones form there?

2. During which months does summer occur in the southern hemisphere?

3. Why is there a temperature difference between the equator and the poles?

Putting a spin on tropical cyclones

Two characteristic features of tropical cyclones are their spiral shape and the curved path they follow. Both are controlled by a phenomenon called the Coriolis effect.

Hurricane Floyd approaches the Florida coast in September, 1999

The Coriolis effect

Each day, Earth makes one full rotation on its axis. To complete this trip, a point at the equator must travel more than 40,000 kilometers in 24 hours—a speed of about 1,670 km/hr (1,035 mph)! As you approach the poles, the distance a point must travel to complete a rotation decreases. Thus, the speed at which the surface is moving also decreases. For example, a point at 45°N travels only about 28,000 km per day, or about 1170 km/hr (727 mph). At the poles themselves, the speed of the surface is essentially zero.

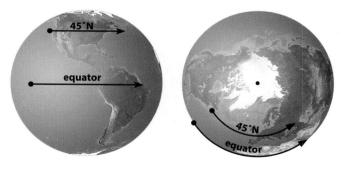

Air near the surface travels at about the same speed as the ground below it. When the sun heats air near the equator, it rises and begins moving toward the pole. As it moves poleward, the speed of the surface below decreases. The air moves faster than the surface, and appears to curve or deflect in the direction of Earth's rotation, the East. The air appears to veer to its right in the northern hemisphere and to its left in the southern hemisphere.

At the poles air cools, sinks, and spreads out toward the equator. Since the air has no rotational speed, it "lags behind" the ground beneath it, and appears to deflect toward the West. Again, the air appears to veer to its right in the northern hemisphere and to its left in the southern hemisphere.

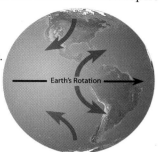

Imagine yourself in a hot air balloon traveling along with the air, following the paths shown by the arrows. As you look ahead in the direction of travel, your balloon would seem to curve to the right in the northern hemisphere and to the left in the southern hemisphere.

For more complex reasons, air moving due east or west follows the same pattern, deflecting to the right in the northern hemisphere and to the left in the southern hemisphere

Coriolis effect and latitude

You've seen that as you move north or south on the Earth's surface, the rotational speed of the surface changes. The rate of change is very small near the equator and increases as you approach the poles. Therefore, the strength of the Coriolis effect at the equator is zero, and increases with latitude. Within about 5 degrees of the equator, the effect is so weak that there's not enough rotation to build or sustain tropical cyclones. Tropical cyclones don't cross the equator into the opposite hemisphere, because they can't maintain their rotation without the Coriolis effect, nor can they change the direction of their rotation.

Low pressure systems

Tropical cyclones are low-pressure systems. This means that air flows in *toward* the center of the system. In the **northern hemisphere** the Coriolis effect deflects this air to the right, causing it to spiral inward in a **counterclockwise** direction.

In the **southern hemisphere** the winds veer to their left, spiraling inward in a **clockwise** direction. This spiral motion, in opposite directions in each hemisphere, produces the characteristic shape of tropical cyclones.

Driving storms

Tropical cyclones are embedded in a sea of air that moves according to differences in air pressure and the Coriolis effect. Earth's atmosphere is organized into six global wind belts, shown here.

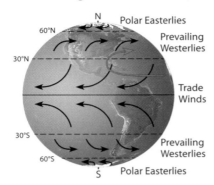

Tropical cyclones form in the trade winds, which blow from east to west. As the Coriolis effect deflects these winds toward the poles, the cyclones may be carried into the westerlies which carry them back toward the east. The storm as a whole follows the same pattern as the air in the global wind belts in which they travel.

- In the northern hemisphere, storms follow **clockwise paths**.
- In the southern hemisphere, storms follow **counterclockwise paths**.

Steering Atlantic hurricanes

The Bermuda High is a semi-permanent area of high pressure centered in the northern Atlantic Ocean. Air moving outward from the high pressure center is deflected to its right creating a clockwise, or anticyclonic circulation.

The four images below show the position of the Bermuda High on the same day, but during different years. You can see how the circulation closely resembles the tracks of many of the Atlantic hurricanes you examined. If the Bermuda High shrinks or shifts eastward, hurricanes stay away from the US coast. Conversely, if the Bermuda High gets larger or shifts westward, hurricanes are more likely to make landfall in the US.

4. As surface winds blow toward the equator in the southern hemisphere, which way are they deflected by the Coriolis effect?

5. In the northern hemisphere, does a tropical cyclone generally follow a clockwise path or a counterclockwise path?

6. Why don't tropical cyclones form near the equator?

7. If the Coriolis effect is strongest near the poles, why don't tropical cyclones form beyond about 35° latitude?

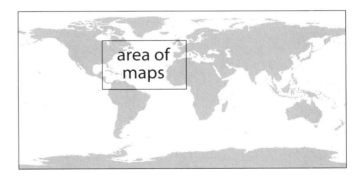

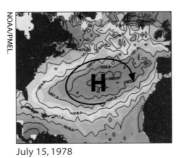
July 15, 1978

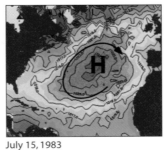
July 15, 1983

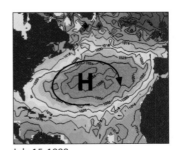
July 15, 1988

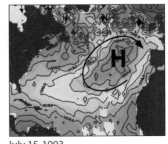
July 15, 1993

Developing the cyclone

Heat, or thermal energy, is a critical factor in forming tropical cyclones. The diagram below illustrates how heat energy from warm ocean water and warm, moist air masses powers these immense storms. The letters refer to the letters in the diagram.

Required ingredients

The process of forming and sustaining a tropical cyclone requires special conditions:

A. Warm ocean waters extending to a depth of at least 50 meters (150 ft) and located at least 500 km from the equator.
B. Converging winds caused by a weak tropical low-pressure system.
C. Warm, moist air that is unstable, meaning that it tends to rise into the atmosphere.
D. The air cools as it rises, eventually reaching a temperature called the *dew point*, where the water vapor condenses into droplets. In addition to creating rain, this process releases heat, called the **latent heat of condensation**.
E. The released heat warms the surrounding air, creating stronger updrafts that increase the convergence of warm, moist air at the surface. This cycle continues until a tropical cyclone develops or something disrupts the process.

Vertical wind shear

Even with all of these ingredients present, there are conditions that can keep tropical cyclones from forming or can cut off their energy source after they have formed. One such atmospheric condition is called *vertical wind shear*.

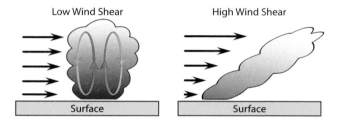

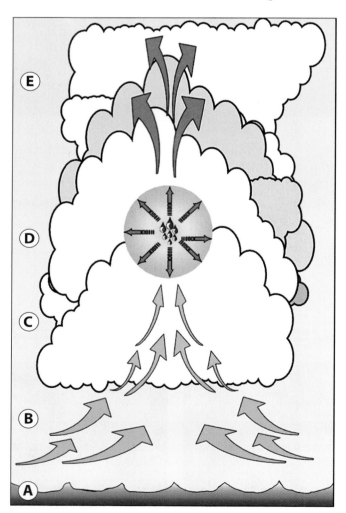

Vertical wind shear is the change in wind speed and/or direction with height. High vertical wind shear disrupts the strong convection by spreading the latent heat released by the condensing water vapor over a wider area.

To form or maintain a tropical cyclone, the vertical wind shear between the surface and the upper troposphere must be less than 37 km/hr (23 mph).

8. How does heat energy power a tropical cyclone? (Refer to diagram at left.)

9. If surface winds are blowing eastward at 15 km/hr and winds in the upper troposphere are blowing westward at 30 km/hr, will there be enough vertical wind shear to prevent a tropical cyclone from forming? (Remember to add wind speeds blowing in opposite directions!)

Activity 1.4

Powering tropical cyclones

Energy from the sun drives Earth's weather. Solar radiation travels through space and Earth's atmosphere and is absorbed by Earth's surface. As the surface heats up, it warms the air above the surface. Our weather is complex for two reasons: Earth's surface doesn't heat up evenly, and the planet is spinning.

Energy for tropical cyclones

So far, you've seen that tropical cyclones form only within a limited range of latitudes and that they usually form during the summer and early fall in the northern and southern hemispheres. This seasonal pattern appears to be related to the warming of Earth's surface by the sun. What temperature range do tropical cyclones form in, and is there some minimum temperature below which tropical cyclones don't form? In this part of the activity, you will try to answer these and other questions.

▸ Launch the ArcView GIS application, then locate and open the **cyclones.apr** project file.

▸ Open the **Powering Tropical Cyclones** view.

▸ Turn on the **August SST (C)** theme.

To turn a theme on or off, click its checkbox in the Table of Contents.

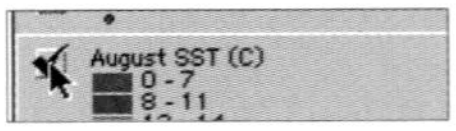

This theme shows the average sea surface temperature (SST) in degrees Celsius (°C) for the month of August, the warmest summer month in the northern hemisphere.

Working with temperatures in Celsius instead of Fahrenheit can be confusing at first. To get a better feeling for the temperatures involved, you will convert two temperatures from Fahrenheit to Celsius.

Temperature Conversions

$°C = (5/9) \times (°F - 32)$

$°F = ((9/5) \times °C) + 32$

1. Use the conversion $°C = (5/9) \times (°F - 32)$ to convert 70°F and 80°F to degrees Celsius (°C). Round to the nearest degree.

   70°F = _____ °C   80°F = _____ °C

Read the legend to answer the following questions about the sea surface temperature.

2. What color represents the warmest water? What is the temperature of the warmest water?

Reading coordinates: The cursor's longitude and latitude are displayed on the right end of the tool bar.

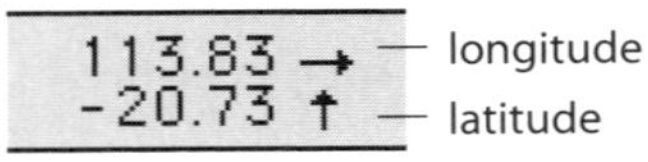

3. In August, water with a temperature of 27–28°C is found as far north as _____ °N and as far south as _____ °S latitude.

▸ Turn off the **August SST (C)** theme and turn on the **February SST (C)** theme.

4. In February, water with a temperature of 27–28°C is found as far north as _____ °N and as far south as _____ °S latitude.

Exploring Tropical Cyclones
Unit 1 - Recipe for a Cyclone

Earth's seasons

Dates	Hemisphere	
	Northern	Southern
Dec 21 - Mar 19	Winter	Summer
Mar 20 - Jun 20	Spring	Fall
Jun 21 - Sep 21	Summer	Winter
Sep 22 - Dec 20	Fall	Spring

To turn a theme on or off, click its checkbox in the Table of Contents.

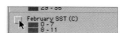

To activate a theme, click on its name in the Table of Contents.

Creating a unique value legend

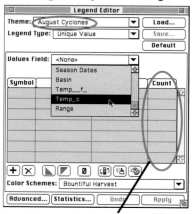

Read the number of cyclones that formed at each temperature in the **Count** column of the Legend Editor. Ignore the **9999** value - it was used if the temperature of formation was uncertain.

Sea surface temperature and the seasons

To see how the sea surface temperature changes throughout the year, a series of SST maps have been assembled into a movie.

▶ To view the movie, click the Media Viewer button 🎬 and choose **SST Movie** from the list. You can step through the movie a frame at a time using the left and right arrow keys on the keyboard.

Click the play button and watch the movie several times. The colors represent the sea surface temperature averaged over a nine year period.

5. How does the shift in sea surface temperature reflect the seasons in the northern and southern hemispheres?

▶ Close the movie window when you are finished viewing the movie.

▶ Turn off the **February SST (C)** theme.

Searching for a threshold temperature

Is there some minimum temperature needed to form tropical cyclones? To find out, you will use the Legend Editor to classify tropical cyclones according to the temperature of the ocean surface over which they formed.

▶ Turn on and activate the **Tropical Cyclones (Aug)** theme.

This theme shows the starting location of every tropical cyclone that occurred during the month of August from 1950 to 2000. The theme's data table also includes the average August sea surface temperature for that location, in degrees Celsius.

▶ To classify the August tropical cyclones according to the average temperature of the ocean surface over which they formed:

- Double-click the **Tropical Cyclones (Aug)** theme in the Table of Contents to open the Legend Editor window. (see left)
- Set the **Legend Type** to *Unique Value* and the **Values Field** to *Temp_C*.
- Read the number of cyclones that formed at each temperature in the Count column.

6. Record your results in the table below.

Temperature (° C) ➤	25	26	27	28	29	30
# of Tropical Cyclones – August						N/A
# of Tropical Cyclones – February						

▶ Close the Legend Editor window.
▶ Turn off the **Tropical Cyclones (Aug)** theme and turn on and activate the **Tropical Cyclones (Feb)** theme.
▶ Repeat this process to classify the February tropical cyclones according to temperature and record the results in the table.

20 Powering tropical cyclones

7. Plot the number of tropical cyclones versus temperature (°C) for August and for February on the graph below. Use a solid line for August and a dashed line for February.

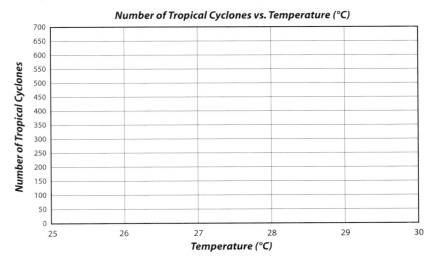

8. Using the graph, at what temperature do tropical cyclones first appear in the northern (August) and southern (February) hemispheres?

9. At what temperature do they occur most frequently?

Draw a vertical line on the graph at 26.7°C. This is the temperature that experts say is needed for a tropical cyclone to form.

10. How well do the data you graphed agree with the experts?

Hint for question 11

Look at one of the sea surface temperature themes. What would a graph of the sea surface temperature versus area look like?

11. Logically, the warmer the water, the more tropical cyclones you should have. Why do you think the number of tropical cyclones on the graph actually decreases for temperatures above 28°C?

What do the scientists say?

For a detailed though somewhat technical discussion of the effects of global warming on tropical cyclone formation, visit Christopher Landsea's *FAQ: Hurricanes, Typhoons, and Tropical Cyclones* website at:

http://www.aoml.noaa.gov/hrd/tcfaq/tcfaqG.html#G3

12. If global warming is a real phenomenon, and ocean temperatures increase worldwide, how do you think this could affect the frequency (how often), latitude range, and intensity of tropical cyclones. Justify your answer.

Activity 1.5 — Solving the cyclone puzzle

In Activity 1.2, you discovered a mysterious situation. Tropical cyclones form in the lower latitudes of most of the world's oceans, yet are virtually absent from the Southern Atlantic and Southeastern Pacific Oceans.

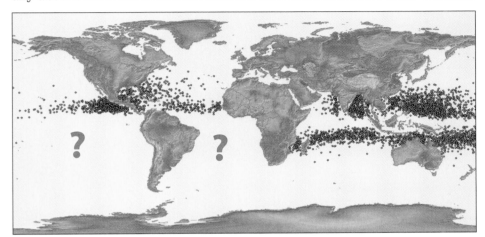

Tropical cyclone ingredients

- A **weak tropical low-pressure system** that starts the whole process. This system causes air to flow into the low-pressure center, creating wind. (Activity 1.3)
- A stable energy source to maintain convection. This energy source is a large body of **warm (27°C or warmer) ocean water**. The water must extend to a depth of about 50 meters to support the storm. As the water evaporates, it intensifies the convection started by the low-pressure system. (Activities 1.3 and 1.4)
- To maintain strong convection, the region should have **low vertical wind shear**—less than 37 km/hr or 23 mph from the surface to the upper troposphere. (Activity 1.3)
- Finally, the storm must form about **500 km or more from the equator**, where the Coriolis effect is strong enough to cause the deflection of horizontal winds. (Activities 1.2 and 1.3)

A process of elimination

One way to solve a problem is to use a *process of elimination* to find the solution. By identifying the ingredients that are *not* the cause of the problem, you may be able to narrow the choices of the correct cause down to just one or two possible "culprits."

Tropical cyclone checklist

In this unit, you have examined specific conditions or "ingredients" that must be present for tropical cyclones to form and grow. The four key ingredients are described at left. Based on everything you have read and observed in this unit, see if you can figure out why tropical cyclones almost never form in these two areas.

1. Complete this tropical cyclone formation checklist for the Southern Atlantic and Southeastern Pacific Oceans.

	yes	no	don't know
a. Do weak, tropical low-pressure systems form there?	☐	☐	☐
b. Does the ocean surface there reach temperatures of 27°C or warmer?	☐	☐	☐
c. Do the regions have low vertical wind shear?	☐	☐	☐
d. Are parts of the region more than 500 km away from the equator?	☐	☐	☐

2. Which ingredient(s) do you think is (are) most likely missing in these cyclone-free areas? Explain how this prevents tropical cyclones from forming.

3. What would you do to find out if your answer to this puzzle is right? What data would you like to collect and add to your map?

Unit 2
The Life of a Cyclone

In this unit, you will learn...

- *the life story of a famous North Atlantic hurricane,*
- *the stages in the life of a tropical cyclone,*
- *how, where, and why tropical cyclones die,*
- *how scientists track tropical storms, and*
- *how to determine the intensity of a tropical storm.*

Three satellite images superimposed on the same map show the development of Hurricane Georges on September 25, 1998.

Exploring Tropical Cyclones Unit 2 - The Life of a Cyclone

Activity 2.1

Observing tropical cyclones

In Unit 1.1, you read an eyewitness account of the 1900 Galveston Hurricane, the worst natural disaster in United States' history. Why was the Galveston hurricane so destructive? Part of the answer is that the weather forecasters in Galveston were forced to make predictions based almost exclusively on direct observation. Watching the skies, the tides, and the changing atmospheric pressure finally told them that a hurricane was coming. Unfortunately, by then the storm was almost upon Galveston. The forecasters had little time to warn the residents to evacuate or seek shelter.

Today's weather forecasters use sophisticated tools to locate and track tropical cyclones well before they make landfall. By modeling tropical cyclones on computers, researchers can predict the direction storms will move more accurately. Forecasters use modern tools to observe and measure cyclones from a distance using indirect observations. To learn about both direct and indirect methods of observing tropical cyclones, point your Web browser to the NOAA site below.

http://hurricanes.noaa.gov/prepare/observation.htm

On the NOAA Web page, click on each of the sensors to learn more about it. Find out how the sensor helps scientists monitor these dangerous storms. Fill in the blank next to each sensor with the sensor type and tell whether it is a direct or indirect method of observation.

Who is NOAA?

NOAA is the National Oceanic and Atmospheric Administration, a government agency that conducts research and gathers data about the global oceans, atmosphere, space, and sun, and applies this knowledge to science and service that touch the lives of all Americans.

NOAA warns of dangerous weather, charts our seas and skies, guides our use and protection of ocean and coastal resources, and conducts research to improve our understanding and stewardship of the environment.

Part of the US Department of Commerce, NOAA provides services through five major organizations.

- National Weather Service
- National Ocean Service
- National Marine Fisheries Service
- National Environmental Satellite, Data and Information Service
- NOAA Research and special program units

1. What are the two main types of satellite imagery used in tropical cyclone forecasting?

2. Radiosondes and dropsondes help us understand the structure of tropical cyclones by measuring and transmitting information about which atmospheric properties?

3. What is the most direct method of measuring the wind speeds within a tropical cyclone?

4. Which two important properties of tropical cyclones are measured using radar?

Exploring Tropical Cyclones — Unit 2 - The Life of a Cyclone

Activity 2.2

Tracking Hurricane Georges

Tropical cyclones don't start out fully developed. Like people, who grow from infant to toddler to teenager to adult, tropical cyclones also experience stages of growth. Unfortunately, instead of "mellowing out", hurricanes get nastier and generally more dangerous with age. This activity focuses on the development of a single hurricane, Georges (pronounced *zhorzh*), that wreaked havoc on Atlantic and Caribbean coastal communities in September 1998.

▶ Launch the ArcView GIS application, then locate and open the **cyclones.apr** project file.

▶ Open the **Tracking Georges** view.

This view shows the path, or *track*, of Hurricane Georges as it swept across the Atlantic and the Caribbean. Hurricane Georges ranks as the 19th most deadly tropical cyclone in history. The large number of casualties occurred as Georges passed directly over several Caribbean islands.

An eye in the sky

This view's table of contents contains seven themes, labeled by date and time, that show satellite images of Hurricane Georges at different points in its journey across the Atlantic Ocean.

▶ Beginning with the 9/15 image, turn on each of the seven hurricane image themes in order.

The hurricane's clouds weren't really bright colors like these. The colors are used to show the temperature of the cloud tops recorded by the satellite's infrared sensor. Gray and white represent the warmest temperatures, while green, yellow, and red are the coldest. Since temperature

To turn a theme on or off, click its checkbox in the Table of Contents.

Georges in false color

In this false color satellite image of Hurricane Georges, the highest (coldest) cloudtops are shown in orange and yellow. These are found where convection is the strongest.

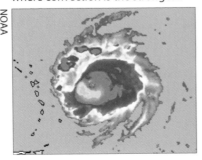

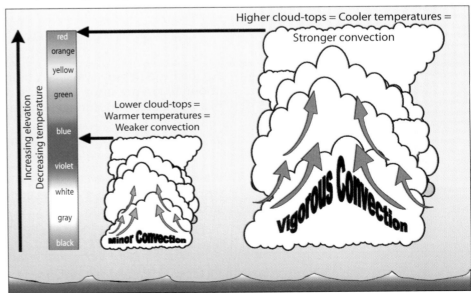

Tracking Hurricane Georges 29

decreases with altitude, colder temperatures indicate higher cloud tops. As convection in a storm intensifies, the clouds extend higher into the atmosphere. Thus, the cloud temperatures tell us about the strength of the convection, which in turn indicates the energy of the hurricane.

1. On what date did Hurricane Georges make landfall in the US mainland? How many days did it take Hurricane Georges to travel across the Atlantic Ocean before landfall?

Universal Time (UT)

The times are listed in 24-hour Universal Time format. Abbreviated UT, this is the time at 0° longitude, or the Prime Meridian. Therefore, 23:46 represents 11:46 pm UT.

The US Atlantic Coast is 5 hours west of the Prime Meridian, so subtract 5 hours from Universal Time to get local time for the Atlantic Coast. (In this example, 6:46 pm ET.)

2. Convert the number of days from question 1 into hours.

3. In what direction did Georges travel as it crossed the ocean?

Now you will measure the distance Georges traveled and use it to find Georges' average rate of travel, or speed.

▶ Using the Measure tool 📇, click on the 9/15 storm center. Click on each of the other storm centers, in order, then double-click on the storm center at landfall (9/27). The total distance Georges traveled (Length) is given in the status bar. *(Your Length will be different.)*

4. How far did Georges travel, in kilometers?

5. Use this distance and the number of hours you calculated in question 2 to calculate Georges' average speed in km/hr as it crossed the Atlantic Ocean. (speed = distance / time)

6. An average person walks at about 6.5 km per hour, and runs at about 10 km per hour. Compare these rates with the speed you just calculated for Hurricane Georges.

7. If a major tropical cyclone is spotted 500 km offshore, how much warning time might a coastal community have before the storm strikes?

Exploring Tropical Cyclones | Unit 2 - The Life of a Cyclone

On the path of a killer

Next you will learn about the countries affected by Georges as it traveled across the Atlantic. Look at the **Storm Track** theme, which shows the path of the center of the hurricane as it crossed the Atlantic.

To turn a theme on or off, click its checkbox in the Table of Contents.

To activate a theme, click on its name in the Table of Contents.

▸ Turn the Georges images on and off as needed to better see the track.
▸ Use the Identify tool [i] to learn the names of countries, islands, and states along Georges' path:
  • Activate the **Countries** theme.
  • Using the Identify tool [i], click on a country or island, and read the **Name** in the Identify Results window.
  • If you need to identify US states, activate the **States** theme before using the Identify tool [i].

8. Identify some of the landmarks associated with Hurricane Georges as it made its way across the Atlantic Ocean:
   a. Hurricane Georges formed just to the south of this group of islands off of the coast of Africa: _____.
   b. Georges passed directly over three large Caribbean island countries and one US territory. These are _____, _____, _____, and _____.
   c. The first US state affected by Georges' heavy rainfall and strong winds was _____.
   d. The US state where the eye of Hurricane Georges first made landfall was _____.

Hurricane formation

▸ Close the **Tracking Georges** view and open the **Formation and Movement** view.
▸ Turn on the **Formation Points** theme.

Each point in this theme indicates where a storm system began showing the characteristic features of a tropical cyclone. Notice that the points are not evenly distributed.

9. On the map below, identify areas where hurricane formation in the Atlantic appears to be concentrated.

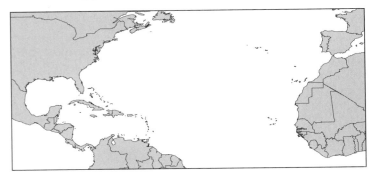

▸ Turn on the **End Points** theme.

To turn a theme on or off, click its checkbox in the Table of Contents.

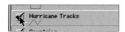

Don't be fooled!
Remember from your look at Georges that each hurricane track represents the path traveled by a large and dangerous storm that affects an area far larger than the thin lines you see.

To activate a theme, click on its name in the Table of Contents.

Look for areas where hurricanes appear to lose strength and die.

10. Identify and label areas on the map on page 31 where hurricanes tend to die out.

▶ Turn on the **Hurricane Tracks** theme.

The tangled lines are the paths traveled by Atlantic hurricanes from 1950 to 1999. Each path was defined by recording the location of the storm's center at regular intervals and then connecting the dots. Before the existence of weather satellites, storm locations were determined by aircraft and reports from ships at sea. Next, you will find out the number of hurricanes that have crossed the Atlantic over the past fifty years.

How many hurricanes?

Every hurricane is a unique event with its own "personality." This may be one reason why it seems fitting to give hurricanes human names. Still, most hurricanes go through similar life cycles.

▶ Determine how many hurricanes occurred between 1950 and 2000.
  - Activate the **Hurricane Tracks** theme.
  - Click the Open Theme Table button [▦].
  - Read the total number of hurricanes from the bar above the table window:

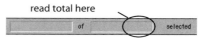

11. How many Atlantic Basin hurricanes occurred between 1950 and 2000?

12. What is the average number of hurricanes that occur each year in the Atlantic Ocean basin? (total hurricanes / years)

Global wind patterns and hurricane tracks

What are "prevailing winds"?
Prevailing wind means the most common wind direction, averaged over a long period of time and over a broad area. In general, winds over the US mainland blow from the southwest, and are called the *prevailing westerlies*.

Reading coordinates: The cursor's longitude and latitude are displayed on the right end of the tool bar.

```
113.83 →   longitude
-20.73 ↑   latitude
```

Next you will look at global wind patterns and determine their influence on hurricane movement in the Atlantic Ocean and the Gulf of Mexico.

▶ Turn on the **Global Wind Patterns** theme. The arrows indicate the direction of prevailing winds over the oceans.

13. In what direction do the prevailing winds over the Atlantic Ocean appear to be moving near the equator?

14. In what direction do the prevailing winds over the Atlantic Ocean appear to be moving near 40°N latitude?

15. Do the prevailing winds over the Atlantic appear to be rotating? If so, are they rotating clockwise or counterclockwise?

▶ Turn off the **Global Wind Patterns** theme.

Exploring Tropical Cyclones Unit 2 - The Life of a Cyclone

To activate a theme, click on its name in the Table of Contents.

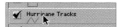

- ▸ Activate the **Hurricane Tracks** theme.
- ▸ Using the Select Feature tool, click on one of the hurricane tracks. One or more hurricane paths will be highlighted.
- ▸ Repeat this several times at random locations and try to identify any general patterns in hurricane movement.

16. Draw the general pattern(s) of hurricane movement on the map below. Label the direction they move once formed.

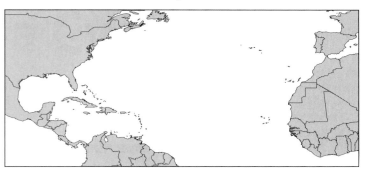

- ▸ Choose **View ▸ Full Extent** to see the entire map.

To turn a theme on or off, click its checkbox in the Table of Contents.

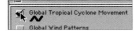

- ▸ Turn on the **Global Tropical Cyclone Movement** theme.

17. Describe the relationship between the paths of Atlantic tropical cyclones and global wind patterns.

Tracking Hurricane Georges 33

Activity 2.3 — Classifying tropical cyclones

An average tropical cyclone lasts about 12 days. If conditions are favorable, it may progress through four stages of development.

1 - Tropical disturbance

A tropical cyclone begins as a low-pressure area with warm, moist air rising from the ocean surface. As the moisture condenses, clouds form and precipitation begins. If the system maintains itself for 24 hours, it is called a **tropical disturbance**. Tropical disturbances have clouds, precipitation, and wind speeds up to 36 km/hr (22 mph), but very little rotation.

2 - Tropical depression

As convection intensifies and the surface pressure continues to decrease, the tropical disturbance begins to rotate. When the winds reach speeds over 36 km/hr (22 mph), the system is classified as a **tropical depression**.

3 - Tropical storm

If a tropical depression intensifies, with wind speeds increasing to 63 km/hr (39 mph) or higher, it is called a **tropical storm** and is assigned a name. The "eye" of the storm may become visible, and the storm begins to resemble a tropical cyclone in many ways.

4 - Tropical cyclone

When wind speeds exceed 117 km/hr (73 mph), the storm is classified as a **tropical cyclone** or **hurricane**. Distinct bands of thunderstorms rotate around the "eye" of the storm. The eye is an area of calm surrounded by the "eye wall," where the winds reach their maximum speed.

Tropical cyclone structure

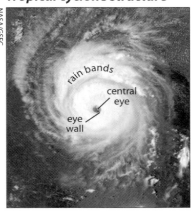

Hurricane Andrew displays a distinct eye and rain bands just before landfall. The eye is a region of relatively calm, clear air that forms as rising air from the thunderstorms converges and sinks in the center of the storm. The eye is surrounded by a ring of tall thunderstorms, the eye wall, where the strongest winds and rain are found. Spiraling inward toward the eye wall are long bands of thunderstorms, the rain bands.

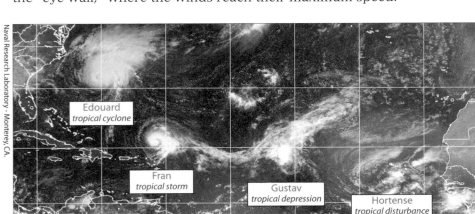

This unique satellite image taken on August 31, 1996 shows systems at each of the four stages described. Hurricane Edouard, Tropical Storm Fran, Tropical Storm Gustav (soon to be downgraded to a tropical depression), and Tropical Disturbance #8 (which eventually developed into Hurricane Hortense).

Exploring Tropical Cyclones — Unit 2 - The Life of a Cyclone

1. If a tropical storm is "downgraded" to a tropical depression, what does this mean about the wind speeds?

2. Fewer than 10% of tropical disturbances evolve into tropical cyclones. Based on what you've learned about tropical cyclones, what causes this percentage to be so low?

Intensity scales

Forecasters responsible for monitoring tropical storms have developed scales to describe a storm's potential for destruction. The US uses the Saffir-Simpson hurricane intensity scale, shown below. This scale ranks hurricanes in five categories from 1, the weakest, to 5, the strongest.

What are knots?

Knots are a measure of speed used in air and water travel.

One knot is equal to:
- 1 nautical mile per hour.
  (A nautical mile is 1/60 of one degree of latitude or 1.15 US statute "land" miles.)
- 1.852 km/hr
- 1.15 mph

The word "knot" comes from an early type of ship's speedometer—a rope with regularly-spaced knots and a weight on the end to make it drag in the water. The pilot threw the end overboard, and counted the number of knots that ran out while a sand glass emptied. This gave the speed of the ship in nautical miles per hour.

Saffir-Simpson Category	Maximum sustained wind speed			Min. Surface Pressure	Storm Surge	
	mph	km/hr	knots	mb	ft	m
Tropical Depression	< 39	< 63	< 34	---	---	---
Tropical Storm	39 - 73	63 - 117	34 - 63	---	---	---
H1	74 - 95	118 - 152	64 - 82	> 980	3 - 5	1.0 - 1.7
H2	96 - 110	153 - 176	83 - 95	979 - 965	6 - 8	1.8 - 2.6
H3	111 - 130	177 - 209	96 - 113	964 - 945	9 - 12	2.7 - 3.8
H4	131 - 155	210 - 250	114 - 135	944 - 920	13 - 18	3.9 - 5.6
H5	156+	251+	136+	<920	19+	5.7+

The Saffir-Simpson Hurricane Intensity Scale categorizes tropical cyclones based on their maximum sustained surface wind speed and lists typical surface pressures and surge heights for each category.

Why rank hurricanes?

The categories of the Saffir-Simpson and similar scales used around the world estimate the potential for flooding and other damage from tropical storms. Wind speed is related to the surface pressure in the center of the hurricane. The maximum wind speed and air pressure reflect the storm's energy and destructive potential. Category 3–5 hurricanes are considered to be major, and are capable of causing tremendous damage.

Measuring wind speed

The most meaningful way to measure wind speed is to calculate the average surface wind speed over a period of time. It is then reported as the *maximum sustained surface wind speed*.

The main difference between intensity scales used by the US and other countries is the length of time over which the wind speed is averaged. The World Meteorology Organization recommends using a 10-minute average, while US forecasters continue to use a 1-minute average.

Storm surge

Storm surge
For Atlantic storms, the surge is usually located in the right front quadrant, where the storm's rotation and its forward motion add to produce the strongest winds.

Hurricane Andrew approaches the Louisiana coast on August 25, 1992.

Storm surge is a "mound" of water piled up on the leading edge of a tropical storm by strong winds blowing across the ocean surface. As the surge washes ashore, it temporarily raises sea level over wide areas of the shoreline. Storm surge can be the most destructive aspect of a tropical storm.

Storm intensity and damage

The Saffir-Simpson Hurricane Scale provides an estimate of the damage potential of hurricanes based on their wind speeds and storm surge heights.

Use the Saffir-Simpson chart to answer the following questions:

3. As a storm grows from a tropical depression to a Category 5 hurricane, how do the surface pressure and the wind speed change?

4. Storm surge adds to the normal tides for a section of coastline. How would the time of day affect the severity of the storm surge striking coastal areas?

5. How high would sea level be if an H3 storm made landfall at high tide and the normal tides range ±2 meters from mean sea level?

6. How high would sea level be if the same H3 storm made landfall at low tide?

The chart below shows damage statistics for Atlantic hurricanes from 1950 to 2000. It includes the average damage per storm and the total number of storms in each category for which damage figures are known.

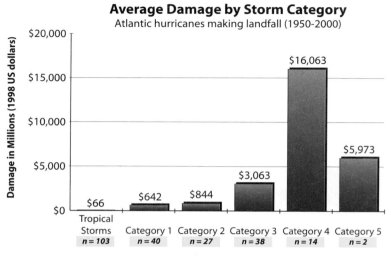

Use the chart to answer the following questions.

7. In general, what happens to the number of hurricanes as the category increases?

8. In general, what happens to the damage caused by hurricanes as their intensity increases?

9. About how many times more destructive, in terms of dollar damage, is an average H4 hurricane than an average H2 hurricane?

10. What percentage of the Atlantic hurricanes making landfall from 1950 to 2000 were Category 4 or 5?

Logically, deaths and property damage should both increase with hurricane intensity. According to the chart, an average H4 hurricane causes more damage than an average H5. This doesn't appear to make sense, since H5s are the most powerful hurricanes.

11. Describe a situation where an H5 hurricane could strike land but cause little damage to structures and no deaths.

Activity 2.4 — Monitoring cyclone growth

In an earlier activity, you examined the path of Hurricane Georges across the Atlantic Ocean. In this activity, you will use measurements of the wind speed to classify the intensity of the hurricane at each point along its path.

▶ Launch the ArcView GIS application, then locate and open the **cyclones.apr** project file. Open the **Tracking Georges** view.
▶ Turn on the 9/15 Georges theme. The other Georges satellite image themes should be turned off.
▶ Activate the **Storm Centers** theme.
▶ Use the Identify tool ⓘ to get information on the storm center in the middle of the Georges 9/15 image.

To turn a theme on or off, click its checkbox in the Table of Contents.

To activate a theme, click on its name in the Table of Contents.

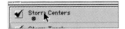

1. Record the pressure in millibars (mB) and wind speed in kilometers per hour for September 15th in the table below.
2. Use the pressure and wind speed to find the Saffir-Simpson category on this date and enter it in the table.

Early in the development of a tropical cyclone, the storm begins to rotate. Look carefully at the image for evidence of rain bands, a central eye, or any other visual clue that the storm has begun to spin.

3. In the table, write either N (none), F (faint), or S (strong) for the storm's **Rotation**.

Tropical cyclone structure

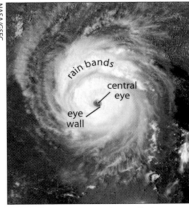

Hurricane Andrew displays a distinct eye and rain bands just before landfall.

		Hurricane Georges image date					
characteristic	how to find	9/15	9/17	9/18	9/20	9/23	9/27
Pressure	use Identify tool ⓘ						
Wind Speed	use Identify tool ⓘ						
Category	Saffir-Simpson Scale						
Rotation	look at Georges images						

▶ Repeat this process to turn on the other Georges satellite image themes and complete the table for the remaining dates.

Answer these questions based on the table and your observations:

4. When did Georges first reach hurricane status? How did its appearance change from the previous image?

5. When was Georges at peak intensity?

6. What was Georges' maximum wind speed at its peak intensity?

Using the measure tool

Click the measuring tool on one side of the hurricane, drag across to the opposite side of the storm, and read the distance in the status bar. Double-click to stop measuring.

▶ Use the Measure tool to measure the diameter of Hurricane Georges at its peak, when its wind speeds were the highest. Read the diameter in the status bar.

read distance here (your distance will vary)

7. How wide (in km) was Georges at its peak?

8. How did Georges' structure change after it passed over Haiti and the Dominican Republic?

9. Look at the 9/29 image of Georges. How did the storm's appearance change? What caused this change?

Major hurricanes

The final part of this exercise looks at the life cycles of major hurricanes, using the sixty-six Category 4 and 5 hurricanes that have occurred since 1950. These storms have wind speeds greater than 130 mph and are capable of catastrophic damage.

As hurricanes develop and move across sea and land, their energy, wind speed, and potential for damage change. In this section, you will examine some of these changes and identify where hurricane intensity appears to increase or decrease.

▶ Close the **Tracking Georges** view and open the **Hurricane Intensity** view.

This theme shows the storm tracks for all of the Category 4 and 5 hurricanes since 1950. Changes in the width and color of the tracks indicate changes in the wind speed as the storm progressed from Tropical Storm to a Category 4 or 5.

Now you will examine the relationship of wind speed to storm location in more detail. The hurricanes in each of the five outlined areas have unique characteristics. You will be investigating these characteristics and recording them in the table below.

▶ To identify differences in hurricane wind speed:
- Select the outline of an area by clicking on the edge of the boundary line using the Pointer tool. Handles (small boxes) appear around the boundary to indicate that it has been selected.
- Click the Select Features Using Graphics button. The hurricane tracks that fall within the outline will be highlighted yellow.
- Click the Open Theme Table button to open the theme's attribute table.
- Make the **Wind Speed** field active by clicking on the column heading. The field name should be highlighted dark gray.
- Choose **Field ▶ Statistics** to determine the average (mean) wind speed for the hurricanes selected by the graphic.

Selecting areas

To select an area, click its border using the Pointer tool. Eight "handles" appear when the area is selected. Below, Area 3 is selected.

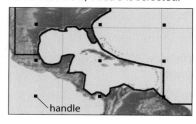

To activate a table field, scroll across and click on its name. Active table fields are highlighted gray.

Area locator

10. Record the average wind speed and whether the area is over land or over water for the area in the following table:

Area	Average wind speed (km/hr)	Over land or over water?
1 - Coast of Africa to Mid-Atlantic		
2 - Mid-Atlantic to US East Coast		
3 - Caribbean and Gulf Coast		
4 - Eastern US		
5 - North Atlantic		

- Repeat this process for the other four areas to complete the table.
- Click the Unselect Features button to deselect the highlighted hurricane tracks.
- Turn on the **August SST (C)** theme.

To turn a theme on or off, click its checkbox in the Table of Contents.

Use the sea surface temperature map and the table to answer these questions about Atlantic hurricanes:

11. Describe the location(s) and, if available, the sea surface temperature where the wind speeds are the highest.

12. Describe the location(s) and, if available, the sea surface temperature where the wind speeds appear to diminish.

13. Briefly describe how the wind speed (and thus the storm energy) changes:
    a. When hurricanes move over warm, tropical water.

    b. When hurricanes move onto land.

    c. When hurricanes move over colder water.

    d. When hurricanes move over the Gulf of Mexico.

Unit 3
Hurricane Hazards

In this unit, you will learn...

- *what makes a hurricane deadly,*
- *which places in the US are the most hurricane-prone,*
- *how hazards differ from risks,*
- *how to determine the likelihood of a hurricane strike in coastal areas,*
- *how hurricanes cause death and destruction, and*
- *who is most at risk from hurricanes in the US.*

The New England Hurricane of 1938 devastated the Harbor View district, southeast of New Bedford, Massachusetts. A storm-shocked husband and wife survey the remains of their home and community. This storm traveled 600 miles in 12 hours, surprising southern New England and causing widespread destruction.

Exploring Tropical Cyclones | **Unit 3 - Hurricane Hazards**

Activity 3.1

Sources of hurricane risk

As you worked your way through **Unit 1 - Recipe for a Cyclone**, you learned where tropical cyclones form, when they form, and how cyclone intensity is related to the presence of warm ocean water. Think for a moment about the dangers that cyclones pose to humans. Certainly a cyclone's high winds are one serious hazard to human life and property, but there are others as well.

Familiar and unfamiliar hazards

In this activity, you will research hazards associated with tropical cyclones. Many of these hazards will be familiar, but some of them may surprise you. Using a Web browser, open the University of Illinois World Weather Project Hurricane Web page.

http://ww2010.atmos.uiuc.edu/(Gh)/guides/mtr/hurr/damg/home.rxml

Additional hurricane hazard information

Another Web site with excellent information about hurricane hazards is sponsored by NOAA.

http://hurricanes.noaa.gov/prepare/title_hazards.htm

If any of these websites are no longer available, visit the SAGUARO Project website for a list of updated addresses.

http://saguaro.geo.arizona.edu

Follow the links on the page to examine each of the hazards associated with tropical cyclones, and then answer the following questions.

1. On which side of a tropical cyclone are wind speeds the highest?

2. Identify the three factors that contribute to a tropical cyclone's storm surge.

3. Which is greater in a tropical cyclone – the risk to people from flooding due to the heavy rainfall, or the danger posed by severe winds?

4. What makes the tornadoes associated with tropical cyclones so dangerous?

5. Which aspect of tropical cyclones is a particular hazard to swimmers both before and after the tropical cyclone occurs?

Activity 3.2

The top ten US hurricanes

Hurricanes cause an average of almost 5 billion dollars in damage per year to the US mainland. Whether or not you live on the Atlantic or Gulf Coasts, you pay some of these costs in the form of higher taxes and insurance premiums. This activity will look at the hurricane risk in the United States, first focusing on the nation's ten most destructive hurricanes and then looking at general risk data for the entire eastern seaboard.

Traits of deadly hurricanes

What do the most deadly hurricanes have in common? To find out, you will look at a map of these devastating storms and search for meaningful patterns.

▸ Launch the ArcView GIS application, then locate and open the **cyclones.apr** project file. Open the **Top 10 US Hurricanes** view.

This view shows the tracks from the ten most damaging hurricanes in American history, from 1900 to the present. The thickness of the lines represents the relative destructiveness of the hurricane. Keep in mind that even the thin lines show hurricanes that were extremely destructive.

1. Look at the top ten hurricane tracks. What similarities do you see in where they formed, and the general shapes of their paths?

What's the difference between hazards and risks?

The terms hazard and risk are sometimes used interchangeably, but they have distinct meanings. A **hazard** is a condition or event that is capable of causing damage or harm. On the other hand, **risk** is the probability that a hazard actually will cause harm. Risk is usually stated as a frequency or percentage of probability of some event occurring over some period of time and for a specified area or location.

For example, in a particular coastal city, storm surge is a hazard. The risk of storm surge for that city might be stated as something like "a 3% chance per year of experiencing a storm surge higher than 4 meters."

▸ To find the average track length of the top ten hurricanes:
- Activate the **Top 10 Tracks** theme.
- Click the Open Theme Table button 🗐 to open the Top 10 Tracks theme table.
- Click on the **Path Length** field label to select that field. The field name should be highlighted gray when it is selected.
- Choose **Field ▸ Statistics** and read the mean value. This is the average track length of the top ten hurricanes.

2. What is the average path length for the top ten destructive Atlantic hurricanes, in km?

Now, compare this to the average path length for all Atlantic hurricanes.
▸ Close the Top 10 Tracks theme table.
▸ Turn on and activate the **Atlantic Hurricanes** theme.
▸ Open the Atlantic Hurricanes theme table and repeat the process described above to find the average path length of all Atlantic hurricanes.

3. What is the average path length for all Atlantic hurricanes, in km?

To activate a theme, click on its name in the Table of Contents. Active themes are indicated by a raised border.

To activate a table field, scroll across and click on its name. Active table fields are highlighted gray.

To turn a theme on or off, click its checkbox in the Table of Contents.

Exploring Tropical Cyclones — Unit 3 - Hurricane Hazards

▶ With the **Path Length** field label still highlighted, click the Sort Ascending button to sort the paths from shortest to longest.

▶ Scroll down the table and look at the relationship between the path length and the other field values (particularly the **Category**.)

4. Describe any relationships you see between the length of a hurricane's path and its intensity.

To turn a theme on or off, click its checkbox in the Table of Contents.

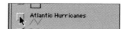

▶ Close the Atlantic Hurricanes theme table and turn off the **Atlantic Hurricanes** theme.

▶ Use the Media Viewer to look at the **Gulf Coast Landfalls** and the **Atlantic Landfalls** maps.

5. Count the number of landfall hits per state and record them in the respective states on the map below. Create a hurricane hazard assessment map by classifying the states into three categories and coloring each state on the map according to its category.

  (1) High Hurricane Hazard (10 or more storms making landfall)
  (2) Moderate Hurricane Hazard (5 to 9 storms making landfall)
  (3) Low Hurricane Hazard (0 to 4 storms making landfall).

Directions

Assign a risk factor based on the number of times a hurricane has crossed a state's boundaries. Assign colors or patterns to each hazard level and color or shade the map and legend accordingly.

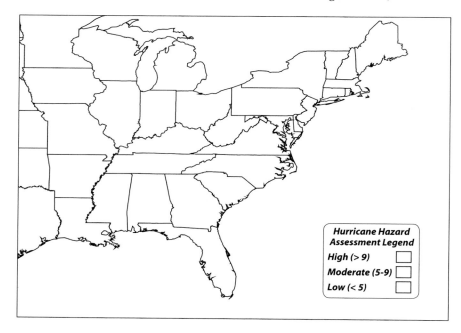

To activate a theme, click on its name in the Table of Contents.

▶ Activate the **Top 10 Tracks** theme.

▶ Click the Open Theme Table button to open the Top 10 Tracks theme table.

48 The top ten US hurricanes

Exploring Tropical Cyclones — Unit 3 - Hurricane Hazards

6. Use the information in the theme table to fill in the following table.

Field ➡	Name	Damage (1998$)	Deaths	Year
Highest dollar damage				
Highest death toll				
Most recent				

Next, sort the hurricanes based on the number of deaths they caused:
▶ Activate the **Deaths** field in the theme table and click the Sort Descending button .

To activate a table field, scroll across and click on its name. Active table fields are highlighted gray.

No deaths in '44?
The casualty figure for the 1944 hurricane is unknown, hence the 0.

Look at the relationship between the number of people killed in each storm (**Deaths**), and the hurricane's date (**Year**). The greatest number of casualties occurred in the Galveston hurricane of 1900, while the fewest occurred with Hurricane Andrew in 1992.

7. Why do you think the number of casualties from major hurricanes is declining, especially since the population of coastal areas is growing?

▶ Close the **Top 10 Tracks** theme table.

Florida – where Hurricanes call "home"

Let's take a closer look at Florida, the state that has been most adversely affected by hurricanes this past century. Although its nickname is the Sunshine State, it's no wonder that the University of Miami sports teams are called the "Hurricanes." Florida has been hit hard by some of the most powerful storms of the 20th century.

To answer the question "Which of the top ten hurricanes crossed Florida?", you will use the Select By Theme operation. The theme containing what you know (**States**) is the selector theme, and the theme containing what you're looking for (**Top 10 Tracks**) is the active theme.

▶ To Select by Theme:
  • Activate the selector theme (**States**).
  • Using the Select Feature tool , click on Florida to select it.
  • Activate the **Top 10 Tracks** theme.
  • Choose **Theme ▶ Select By Theme...**.
  • In the Select By Theme dialog box, select from the popup menus to make the statement read "Select features of the active themes that **Intersect** the selected features of **States**."
  • Click the **New Set** button.

Select By Theme dialog box

The tracks of the top ten hurricanes that crossed Florida should be highlighted.

Exploring Tropical Cyclones — Unit 3 - Hurricane Hazards

Generally, hurricane damage is greatest where the storm first makes landfall, accounting for 90% of the total. To find the extent of the destruction Florida suffered from these hurricanes, you will examine the theme's attribute table:

▶ Open the **Top 10 Tracks** theme table and read the number of selected tracks from the tool bar.

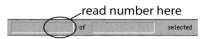
read number here

8. How many of the top ten hurricane tracks crossed Florida?

All of these hurricanes made first landfall in Florida.

To activate a table field, scroll across and click on its name. Active table fields are highlighted gray.

▶ Select the **Damage (1998 $)** field and choose **Field ▶ Statistics**.

9. What is the total damage (**Sum**) from these hurricanes in 1998 dollars?

10. Approximately how much damage did these hurricanes do to Florida in 1998 dollars, based on the 90% rule explained above?

▶ Select the **Deaths** field and choose **Field ▶ Statistics**.

11. What is the total death toll (**Sum**) from these hurricanes?

12. Who do you think pays for these damages? How?

50 The top ten US hurricanes

Activity 3.3
Exploring hurricane hazards

Coastal populations

Humankind has long depended on the sea as a reliable source of food, as a moderating influence on climate, and as a route for trade and travel. However, what on one day is a gentle, benevolent ocean can the next day spawn storms of awesome destruction.

Historical impacts of hurricanes

Historical researchers speculate that Atlantic tropical cyclones are responsible for one-third to one-half million deaths between 1492 and the present. This estimate includes both coastal and offshore losses.

"The total number of ship-related casualties associated with Atlantic tropical cyclones is unknown, but there are clues. Some perspective on the magnitude of ship losses worldwide is gained by realizing that on the coast of England alone there have been a minimum of 250,000 wrecks (Cameron and Farndon, 1984)!

"In fact, to 1825, more than five percent of the vessels in the (West) Indies navigation were lost due to shipwrecks; the biggest part due to bad weather... (Marx 1981)."

Excerpted from http://www.nhc.noaa.gov/pastdeadlytx3.html

Coastal population growth

Americans love the beach, and express that love by living near the coasts in increasing numbers. According to a NOAA report, 54% of the US population lives in coastal counties that account for only 20% of the total US land area. Of those, nearly 60% live on the Atlantic and Gulf Coasts, putting them at risk from hurricanes.

The map shows population growth along the Atlantic and Gulf Coasts from 1960 to 2000. Population has increased dramatically, particularly in Florida, the Carolinas, and Virginia. The more people that live near the ocean, the greater the likelihood that they and their possessions will someday be in the path of a hurricane.

The full report on Coastal Population Growth is available at http://www-orca.nos.noaa.gov/projects/population/population.html

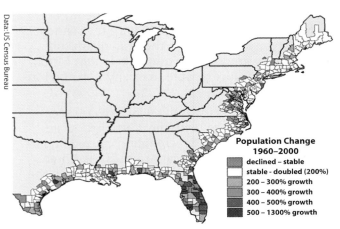

Atlantic and Gulf Coast population growth from 1960–2000.

1. Are ships more vulnerable or less vulnerable to hurricanes today than they were 100 years ago? Explain.

2. Why is the population of coastal areas growing, despite the obvious hurricane risk?

Types of hazards

Storm surge—Galveston Island (1900)

"On September 8, 1900, the greatest natural disaster to ever strike the United States occurred at Galveston, Texas. In the early evening hours of September 8, a hurricane came ashore at Galveston bringing with it a great storm surge that inundated most of Galveston Island and the city of Galveston. As a result, much of the city was destroyed and at least 6,000 people were killed in a few hours time." The storm surge crested at around 5 meters (17 feet), completely submerging the island, whose highest elevation at the time was 2 meters (9 feet).

From the Monthly Weather Review for September 1900. http://www.history.noaa.gov/stories/cline2.html

Landslides and mudflows – Hurricane Mitch (1998)

"In an awesome display of power and destruction, Hurricane Mitch will be remembered as the most deadly hurricane to strike the Western Hemisphere in the last two centuries! The death toll currently is reported as 11,000 with thousands of others missing. Though the final death toll will never be known, it is quite likely that Mitch directly killed more people than any Atlantic hurricane in over 200 years.

"Although the ferocity of the winds decreased during the westward drift, the storm produced enormous amounts of precipitation caused in part by the mountains of Central America. As Mitch's feeder bands swirled into its center from both the Caribbean and the Pacific Ocean to its south, the stage was set for a disaster of epic proportions."

The storm's movement slowed over the volcanic peaks of Central America, and rain fell at the rate of a foot or two per day in some areas. The total rainfall for the storm was as high as 75 inches, or about two meters!

"The resulting floods and mud slides virtually destroyed the entire infrastructure of Honduras and devastated parts of Nicaragua, Guatemala, Belize, and El Salvador. Whole villages and their inhabitants were swept away in the torrents of flood waters and deep mud that came rushing down the mountainsides. Hundreds of thousands of homes were destroyed."

From http://lwf.ncdc.noaa.gov/oa/reports/mitch/mitch.html

Mudflow debris from Hurricane Mitch washed out this bridge in the town of Choluteca, Honduras.

Wind damage – Hurricane Andrew (1992)

"Andrew was a small and ferocious Cape Verde hurricane that wrought unprecedented economic devastation along a path through the northwestern Bahamas, the southern Florida peninsula, and south-central Louisiana. Damage in the United States is estimated to be near 25 billion, making Andrew the most expensive natural disaster in U.S. history.

Damage from Hurricane Andrew depended on the quality and type of construction used. In many neighborhoods, few structures escaped undamaged.

"The tropical cyclone struck southern Dade County, Florida especially hard, with violent winds and storm surges characteristic of a category 4 hurricane, and with a central pressure (922 mB) that is the third lowest this century for a hurricane at landfall in the United States. In Dade County alone, the forces of Andrew resulted in 15 deaths and up to one-quarter million people were left temporarily homeless. An additional 25 lives were lost in Dade County from the indirect effects of Andrew. The direct loss of life seems remarkably low considering the destruction caused by this hurricane."

From http://www.publicaffairs.noaa.gov/andrew92.html

Inland flooding – Hurricane Floyd (1999)

"Floyd brought flooding rains, high winds, and rough seas along a good portion of the Atlantic seaboard from the 14th through the 18th of September. The greatest damage occurred along a path from the eastern Carolinas northeastward into New Jersey and extending to the coast of Maine. Several states had numerous counties declared disaster areas. Flooding caused major problems across the region, and at least 77 deaths have been reported. Damages are estimated to be $1.6 billion in Pitt County, North Carolina alone, and total storm damages may surpass the $6 billion caused by Hurricane Fran in 1996.

"Although Hurricane Floyd reached category 4 intensity in the Bahamas, it weakened to category 2 intensity at landfall in North Carolina. Floyd's large size was a greater problem than its winds, as the heavy rainfall covered a larger area and lasted longer than with a typical category 2 hurricane.

"Approximately 2.6 million people evacuated their homes in Florida, Georgia, and the Carolinas—the largest peacetime evacuation in US history. Ten states were declared major disaster areas as a result of Floyd, including Connecticut, Delaware, Florida, Maryland, New Jersey, New York, North Carolina, Pennsylvania, South Carolina, and Virginia.

"There were several reports from the Bahamas area and northward of wave heights exceeding 50 feet (15 meters). The maximum storm surge was estimated to be 10.3 feet (3.75 meters) on Masonborough Island in New Hanover County, NC."

A summary of Floyd's impact on North Carolina:

- 51 deaths
- 7000 homes destroyed
- 17,000 homes uninhabitable
- 56,000 homes damaged
- most roads east of I-95 flooded
- Tar River crested 24 feet (7.3 meters) above flood stage
- over 1500 people rescued from flooded areas
- over 500,000 customers without electricity
- 10,000 people housed in temporary shelters
- much of Duplin and Greene Counties under water
- severe agricultural damage throughout eastern North Carolina
- Wilmington reported a 24-hour rainfall record of 13.38 inches (34 cm, a 128-year record), and a total of over 19 inches (48 cm) for the event.

View of flooding along Main Street in Tarboro, North Carolina

"Nothing since the Civil War has been as destructive to families here," says H. David Bruton, the state's Secretary of Health and Human Services. "...The recovery process will be much longer than the water-going-down process."

From http://cindi.usgs.gov/ncflood/ncflood.html and http://www.usgs.gov/hurricanes/hurricane-floyd.html

3. What are two ways that tropical cyclones can produce severe flooding?

4. Describe two factors that can significantly increase the rainfall from a tropical cyclone.

5. List and explain several factors that helped make Hurricane Mitch "the most deadly hurricane to strike the Western Hemisphere in the last two centuries."

Exploring Tropical Cyclones Unit 3 - Hurricane Hazards

Activity 3.4 Risk to coastal communities

The devastating effects of hurricanes are controlled by several factors. Clearly, destruction is greater when hurricanes strike cities than when they encounter unpopulated areas.

- Launch the ArcView GIS application, then locate and open the **cyclones.apr** project file. Open the **Top 10 US Hurricanes** view.
- Turn off the **Top 10 Tracks** theme.
- Turn on the **Risk Probability (%)** and the **Major Cities** themes.

To turn a theme on or off, click its checkbox in the Table of Contents.

The **Risk Probability (%)** theme shows the likelihood that an area will be struck by a hurricane in any given year. As you can see, every state along the Atlantic and Gulf Coasts has at least some annual risk for a hurricane strike.

What are the chances?

A probability of 0.20 means that there's a 20% chance that a hurricane will strike in any one year. Over ten years, on average, the area can expect 2 hurricanes in 10 years. (10 years x 20%/year = 200%)

1. What is the range of annual hurricane risk experienced by the states along the Atlantic and Gulf Coasts? (highest and lowest risk)

Warning

Be sure to click directly on the city symbol with the Identify tool when you are finding the risk probability.

- Activate the indicated theme and use the Identify tool 🛈 to determine the population and the annual hurricane risk of the major Atlantic and Gulf Coast cities listed in the table below.

2. Complete the following table, then use the information to rank the cities according the their annual hurricane risk:

Activate this theme	Major Cities	Risk Probability (%)	
City	Population	Annual Risk (%)	Ranking
Boston			
Houston			
Miami			
New Orleans			
New York			
Washington, DC			

3. Using the information in the table, how many hurricanes is Washington D.C. likely to experience in an average decade?

Risk to coastal communities 55

How many people are at risk?

You've looked at the risk in major cities—how many people on the Atlantic and Gulf Coasts are at risk from hurricanes? You will complete this investigation by looking at the total population of all counties with some hurricane risk.

To activate a theme, click on its name in the Table of Contents.

Select By Theme dialog box

▶ Activate the **Risk Probability (%)** theme.
▶ Use the Select Features tool to select all of the risk probability area. The entire area should be highlighted yellow.
▶ Turn on and activate the **Counties** theme.
▶ Choose **Theme ▶ Select By Theme**. Set the pop-up menus to read "Select features of active themes that **Intersect** the selected features of **Risk Probability (%)**," as shown at left and click the **New Set** button.
▶ Open the **Counties** theme attribute table, select the **Pop1999 (est)** field, and choose **Field ▶ Statistics** to determine the total population that is at risk from hurricanes.

4. How many counties (Count) in the US face some risk from hurricanes?

5. What is the total population (Sum) of the counties at risk from hurricanes?

What are the "lower 48 states?"
The "lower 48 states" include all of the states except Alaska and Hawaii. They are also called the Contiguous United States.

▶ Click the Clear All Selected button and calculate statistics on the population field again. This will give the total population of the "lower 48" states.

6. What *percentage* of the population of the "lower 48" states is at risk from hurricanes?

56   Risk to coastal communities

Exploring Tropical Cyclones   **Unit 4 - Hurricanes in the Big Apple**

Unit 4
Hurricanes in the Big Apple

In this unit, you will learn...
- *how to determine hurricane probabilities for coastal cities,*
- *how topography affects storm surge risk,*
- *how government agencies manage emergencies,*
- *who is most at risk from hurricanes and why, and*
- *how to assess risks to the community's infrastructure.*

1997 Landsat 5 image of the New York City area. The hurricane risk to this area is real, and a cause of concern for emergency planners.

Exploring Tropical Cyclones Unit 4 - Hurricanes in the Big Apple

Activity 4.1

Analyzing physical factors

The physical factors of risk

When most people think of cities that are at risk from hurricanes, they think of places like Miami, New Orleans, or maybe Charleston, South Carolina. These cities have had either direct hits or very close calls with major hurricanes in recent memory, and seem to be threatened nearly every year. Few, if any, people associate New York City and hurricanes, believing that New York is too far north, or too far inland, to be in any real danger from a major hurricane.

The reality is that New York City was hit by major hurricanes in 1635, 1815, 1821, and 1893, and escaped with near misses in 1938, 1985 (Gloria), and 1999 (Floyd). Sooner or later, a "Big One" will make a direct hit on New York City, and because of characteristics of the shoreline and the distribution of people in the city, chances are good that the hurricane's impact on the city will be severe.

Probability of a major strike

Hurricanes DO strike New York City, although not as frequently as they hit coastal cities farther south. To begin, you will look at the probability that New York City will be hit by a named storm during any given year. Then you will determine the likelihood of New York City being hit by a major hurricane of H3 or higher.

"Named" storms

When a storm reaches tropical storm intensity, it is assigned a name. The first storm of the season begins with an A, the second with a B, and so on. Each tropical cyclone basin has a standard set of names that are used on a rotating five-year cycle.

▸ Launch the ArcView GIS application, then locate and open the **cyclones.apr** project file.
▸ Open the **Hurricane Probability** view.

This view shows the probability that an area in the Eastern US will get at least one named storm during any given year.

To activate a theme, click on its name in the Table of Contents.

▸ To determine the hurricane probability for New York City:
  • Activate the **Hurricane Landfall Probability** theme.
  • Using the Identify tool ⓘ, click on New York City.
  • In the Identify Results window, read the **Probability** field to find the hurricane probability for New York City.
1. What is the probability, in percent, that New York City will be hit by a named storm in any given year?

Exploring Tropical Cyclones   Unit 4 - Hurricanes in the Big Apple

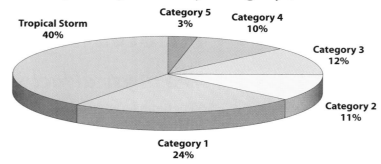

Atlantic Tropical Cyclones by Category (1950–2000)

Use this chart to answer the following questions:

2. What percentage of Atlantic tropical cyclones were minor (i.e., tropical storms, H1, or H2)? What percentage were major hurricanes (H3 or greater)? Record your answers in the table:

Storm Category	% of Atlantic Storms in Category
Minor tropical cyclones: tropical storms, H1 and H2 hurricanes	
Major tropical cyclones: H3, H4, and H5 hurricanes	

The percentage you found in question 1 represents the likelihood that a named storm will hit New York City in any given year. The percentages determined in question 2 tell the percentages of named storms that are minor and major hurricanes.

3. Determine the probability that New York City will be hit by a major hurricane (H3-H5) in any given year. To find this, multiply the percentages (as decimals) obtained in the first two questions:

   _____ (probability of a named storm hitting NYC in any year)
   × _____ (probability of a named storm being a major hurricane)
   = _____ (probability of a major hurricane hitting NYC in any year)

4. Determine the probability that New York City will be hit by a minor hurricane (Tropical Storm-H2) in any given year. To find this, multiply the percentages obtained in the first two questions:

   _____ (probability of a named storm hitting NYC in any year)
   × _____ (probability of a named storm being a minor hurricane)
   = _____ (probability of a minor hurricane hitting NYC in any year)

These probabilities indicate the number of major and minor hurricanes, on average, that could hit New York City each year.

60   Analyzing physical factors

5. How many major hurricanes can New York City expect to get over the course of a century?

_____ (probability of a major hurricane hitting NYC in any year)
× _____ (number of years in a century)
= _____ (number of major hurricanes hitting NYC in a century)

New York City's hurricane history

NYC Hurricane History chart

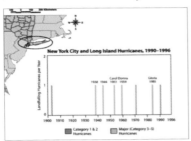

▶ Click the Media Viewer button and choose the **NYC Hurricane History** chart (shown at left).

This chart shows direct hits by hurricanes on New York City and Long Island for most of the 20th century. Examine the chart and think about the probabilities you calculated in questions 4 and 5.

6. How many minor hurricanes affected the New York City area in the 20th century? How many major hurricanes?

   Minor hurricanes = _____

   Major hurricanes = _____

7. Based on the chart and the hurricane probability you determined in question 5, how frequently do you think major hurricanes directly affect New York City? Explain.

Storm Surge

The shape of the coastline can increase the severity of a hurricane's impact on coastal towns and cities by concentrating the effects of the hurricane's storm surge. In this part of the activity, you will examine the Atlantic coast looking for sections of coastline with sharp changes in direction where a hurricane's impact might prove especially severe.

The shape of a coastline can act...

...like a funnel

focusing large volumes of ocean water into smaller bays and inlets,

...like a pot

trapping ocean water in the bays and inlets and keeping it there, or

...like a mirror

reflecting and bouncing the ocean water around in the bays and inlets.

▸ Use the zoom tools to look closely at different parts of the Atlantic seaboard (i.e. from Florida to Maine). Look for sections of the Atlantic coastline that may amplify the effects of the storm surge.

8. Locate three major bays or inlets that may trap water from a storm surge and circle them on the map.

9. Find the one place along the Atlantic Coast where the direction of the shoreline changes from generally north-south to generally east-west. Mark its location on the map.

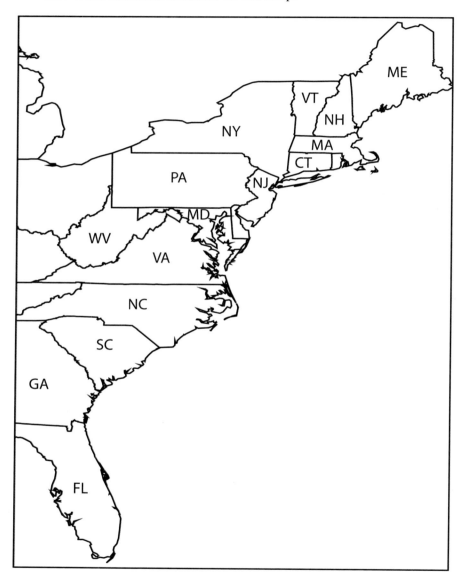

10. How might the shape and orientation of the coastline increase or decrease the impact of the storm surge?

Exploring Tropical Cyclones   **Unit 4 - Hurricanes in the Big Apple**

New York City topography

As you observed, the shape of the New York City coastline is unique. The right angle made by the coastlines of New Jersey and Long Island can funnel storm surge waters into New York Harbor, particularly if the hurricane is approaching directly from the southeast.

Pushing more water into the harbor isn't necessarily a recipe for disaster. If the city were on a plateau 20 meters above mean sea level, a 10 meter surge of water wouldn't have much of an effect. Unfortunately, only part of New York City sits on higher ground, while the rest is very near sea level.

Mean sea level

Mean sea level is the average global sea level. This measure removes the effects of waves and tides, and is used as a reference point for measuring elevations on land.

Now, you will explore New York City topography using a shaded relief map of the city. Using this information, you will get a pretty good idea of the city's flood risk before you look at actual flooding data.

▸ Close the **Hurricane Probability** view and open the **NYC Topography and Demographics** view.

▸ Use the Zoom tools to read the elevation scale on either side of the New York City shaded relief image.

11. What is the range of elevation values for the New York City area?

12. On the map below, circle four general areas that are lower than 5 meters in elevation and four areas that are higher than 10 meters in elevation. Label each high area with an **H** and each low area with an **L**.

Where is this?

The rectangle indicates the area included in the **NYC Topography and Demographics** View.

Analyzing physical factors   63

Storm surge

For Atlantic storms, the surge is usually located in the right front quadrant, where the storm's rotation and its forward motion add to produce the strongest winds.

Hurricane Andrew approaches the Louisiana coast on August 25, 1992.

Storm surge risk

Storm surge is a temporary change in sea level that occurs when strong winds pile ocean water into a bulge ahead of the hurricane. When it reaches the shallow waters at the coastline, the bulge has nowhere to go but upward and inland. Stronger winds produce higher storm surge.

Generally, the higher the storm surge the farther inland the flooding extends. In addition to the surge height, topography plays a significant role in determining the impact on coastal areas. Next, you will look at predicted storm surge data for New York City and explore how topography affects the inundation (flooding) of coastal areas.

▶ Turn on and activate the **NYC Storm Surge** theme. In this theme, darker shades of blue represent higher storm surges.

▶ Choose **Theme ▶ Properties** and click the Query Builder button to open the Theme Properties Query Builder.

▶ Enter the query statement (**[Category] = "Category 1"**) and click **OK** twice to display only the areas flooded by a Category 1 surge.

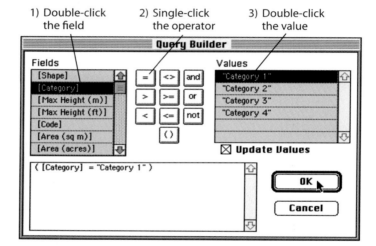

To activate a theme, click on its name in the Table of Contents.

▶ Make sure the **NYC Storm Surge** theme is active and click anywhere within the blue storm surge zone with the Identify tool.

▶ In the Results window, read the maximum storm surge height [Max Height (m)] and the total inundated area [Area (sq km)] and record them in the table.

▶ Repeat this process for Category 2, 3, and 4 storm surges. Choose **Theme ▶ Properties**, open the Query Builder window, and query the next category by changing the number X in the query statement (**[Category] = "Category X"**).

▶ When you have finished measuring all four categories, choose **Theme ▶ Properties** and click the **Clear** button to re-display all four categories.

64   Analyzing physical factors

13. Complete the following table using the information you obtained for each storm surge category.

Storm Category	Surge height (m) [Max Height (m)]	Inundated area (km²) [Area (sq km)]
1		
2		
3		
4		

14. Graph the surge height versus the inundated area on the graph below.

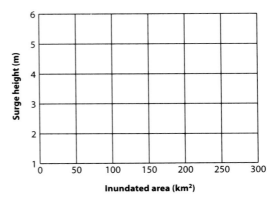

Notice that the inundated area does not increase uniformly from one category to the next. To understand why, you will examine topographic profiles at three different coastal locations.

▶ Turn on the **Topographic Profiles** theme.

The illustrations below show the topographic profiles along the three lines shown on the map.

15. On each profile, draw and label the surge height resulting from category 1, 2, 3, and 4 storm surges at that location. Use the surge heights you recorded in the table in question 13.

a) Profile A

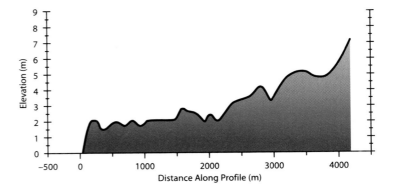

To turn a theme on or off, click its checkbox in the Table of Contents.

Topographic profiles

The lines in this theme mark the locations of *topographic profiles*. A topographic profile is a cross-sectional view of the land surface. Imagine slicing through the Earth, pulling the halves apart, and viewing one half from the side. The result is a topographic profile. To make the land features easier to see, topographic profiles are usually "stretched" vertically—a process called *vertical exaggeration*.

b) Profile B

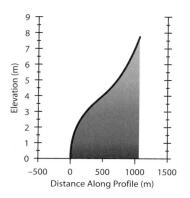

c) Profile C

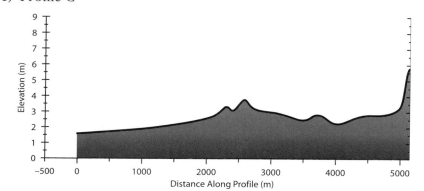

16. Explain how a large increase in the storm surge height can cause only a small increase in the area flooded. Which profile(s) above show examples of this?

17. Explain how a small increase in storm surge height can cause a large increase in the area flooded. Which profile(s) above show examples of this?

Exploring Tropical Cyclones Unit 4 - Hurricanes in the Big Apple

Activity 4.2

Managing emergencies

Most countries, states, counties, large cities, and even some businesses have agencies or offices that plan for what to do in the event of an emergency. Below are the addresses of several websites for emergency management agencies. Choose one or more and explore their sites to find out how they prepare for a natural disaster such as a hurricane, what's done during the disaster, and how the impacts of a disaster are handled once the event has ended. Write a one-page summary of their emergency management plan for hurricanes.

FEMA (Federal Emergency Management Agency)

Main page – http://www.fema.gov
Hurricane page – http://www.fema.gov/fema/trop.htm

State emergency management agencies

Delaware – http://www.state.de.us/dema/NatHaz.htm
Florida – http://www.floridadisaster.org/bpr/emtools/severe/hurricanes.htm
Florida – Internet Evacuation Mapping System (requires a fast Internet connection) – http://mapking.eoconline.org/eoconline/
Georgia – http://www2.state.ga.us/GEMA/
Hawaii – http://www.state.hi.us/dcca/hhrf/hurricanes_hawaii.html
Louisiana – http://www.loep.state.la.us/Default.htm
Maine – http://www.state.me.us/mema/haz_docs/hurricane.htm
Maryland – http://www.mema.state.md.us/hazard_hurricane.html
Massachusetts – http://www.state.ma.us/mema/prepare/hurricane.htm
New York – http://www.nysemo.state.ny.us/NEWSCENTER/Wxaware2000.html
South Carolina – http://www.state.sc.us/epd/
Texas – http://www.txdps.state.tx.us/dem/hurrindx.htm
Virginia – http://www.vdem.state.va.us/season/

New York City Office of Emergency Management

New York City's Hurricane Preparedness Page –
http://www.nyc.gov/html/oem/html/emols/emols_hurricane_application.html

Storm surge information and hazard maps

Escambia County (Florida) animated storm surge –
http://www.escambia-emergency.com/anisurge.html
Escambia County – http://www.escambia-emergency.com/geninfo.asp
NASA – http://www.csc.noaa.gov/products/alabama/htm/hssra.html

---

Has a website moved and left no forwarding address?
The SAGUARO Project maintains a list of current links for this module at
http://saguaro.geo.arizona.edu

Activity 4.3 — Addressing demographic factors

Earlier, you looked at some of the factors that affect hurricane risk and examined the likelihood of a major hurricane striking New York City. So far, however, you have ignored the most important factor—people. This investigation will explore the human element of hurricane risk.

Reverse demographics and hurricane risk

In many American cities, coastal areas are often favored by the wealthy as places to live. In New York City, however, the people in communities along the waterfronts often have lower income and education levels. This situation is a **reverse demographic**, meaning that the social and economic distribution of people is the reverse of what is usually seen.

With respect to hurricanes, this reverse demographic makes an already dangerous situation even worse, since the areas most prone to flooding are inhabited by people who may be:

- less aware of the general dangers of hurricanes;
- less likely to be aware or informed of an approaching hurricane;
- less mobile and therefore less able to quickly evacuate; and
- less likely to be able to afford flood insurance or recover financially if a hurricane damages or destroys their home or workplace.

Demographic factors

Hurricane risk diminishes dramatically as you move away from the shoreline. Most people minimize their personal risk by simply moving inland until the storm passes. Since hurricanes move fairly slowly and are tracked by satellite for days before landfall, people usually have plenty of time to evacuate before the storm strikes. However, for people who don't receive or understand the warning, or who don't have transportation to evacuate, the risk is considerably higher.

In this exercise, you will identify where three specific at-risk populations—the elderly, the poor, and the non-English speaking—live in significant numbers. This will allow emergency planners to effectively focus their resources to assist those most in need.

The analysis is a multi-step process involving both the **NYC Census Data** and the **NYC Storm Surge** themes. *Make sure the correct theme is active for each operation. Otherwise, your data might come from the wrong theme!*

Establish a baseline population

When assessing risk to segments of a population, it is important to first understand the characteristics of the general population as a whole. In this case, you want to know the total population of the area. This provides an important yardstick, or baseline, to which you can compare other measurements.

- Launch the ArcView GIS application, then locate and open the **cyclones.apr** project file.
- Open the **NYC Topography and Demographics** view.
- Turn on and activate the **NYC Census Data** theme.

The gray areas in this theme are called *census block groups*. The US Census Bureau gathers and reports statistics for various sized groups of people. The smallest of these are *census blocks*, areas containing about 70 people. Neighboring census blocks with similar characteristics are combined into *census block groups* of around 500 households each.

- Click the Open Theme Table button 🔲 to open the theme's attribute table.
- Click the **Pop1990** field heading to select it. Choose **Field ▶ Statistics**, and read the SUM. This is the total population for all the census block groups visible on the screen. *NOTE: The view does not show all of New York City. The total population of the city is much larger!*

1. What is the baseline population for the NYC Census Data theme?

- Close the NYC Census Data attribute table.

Identify groups in flooding areas

Having determined the baseline population, the next step is to identify the groups residing in areas that would be flooded by a Category 4 storm surge. You will look at the spatial relationship between the **NYC Storm Surge** theme and the **NYC Census Data** theme, to identify where the two themes overlap. In ArcView, this is done by performing the Select By Theme operation.

- To perform the Select By Theme operation:
  - Turn on and Activate the **NYC Storm Surge** theme.
  - Choose **Theme ▶ Properties**... and click the Query Builder button 🔧.
  - Enter the statement ([Category] = "Category 4") and click **OK** twice. Only the Category 4 storm surge areas will be displayed.
  - Activate the **NYC Census Data** theme.
  - Choose **Theme ▶ Select By Theme** to open the Select by Theme dialog box.
  - In the lower pull-down menu, choose **NYC Storm Surge**.
  - In the upper pull-down menu, choose **Are Completely Within** as the spatial operation.
  - Click the **New Set** button. *(Be patient! This operation may take a couple of minutes on slower computers.)*

When the operation is finished, the census block groups that are completely within the Category 4 storm surge areas are highlighted yellow.

- Open the **NYC Census Data** attribute table. The affected census blocks are highlighted yellow in the table.

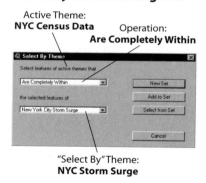

Select By Theme dialog box

Active Theme: **NYC Census Data**
Operation: **Are Completely Within**
"Select By" Theme: **NYC Storm Surge**

Too many / too few census blocks highlighted?

If all or none of the census blocks are highlighted yellow, you probably made a mistake setting up the Select By Theme operation. Redo the operation from the beginning, making sure the correct theme is active at each stage of the process.

- Make sure the **Pop1990** field heading is selected. (It is highlighted gray when selected.)
- Choose **Field ▸ Statistics**, and read the SUM. This is the total population of the census block groups within the areas flooded by a Category 4 storm surge.

2. How many people in New York City would be directly affected by a Category 4 storm surge? What percentage of the baseline population is this?

- Close the NYC Census Data attribute table.

Identify flooding areas with high increased-risk populations

The three increased-risk populations you are trying to identify are the elderly, the poor, and the non-English speaking. Although people in these populations are spread all over the city, you want to know where they are most heavily concentrated. That way, you will be able to focus most of your resources on these areas. The following is the critical concentration (percentage of the population) for each of these groups.

Increased-Risk Population	Category	Critical Concentration
Over 65 years of age	Elderly	$\geq 22\%$
Living below the poverty level	Poor	$\geq 41\%$
Non-English speakers	NoEnglish	$\geq 22\%$

You can assume that the general population will receive the evacuation warnings broadcast by television, radio, and the Internet. Your task is to target specific census block groups for door-to-door evacuation warnings. You will use a query to find the subset of the flooded census blocks that have high increased-risk populations.

- With the **NYC Census Data** theme active, click the Query Builder button.
- Enter the following statement into the query builder dialog box exactly as it appears below:

    ([Elderly] >= 22) OR ([Poor] >= 41) OR ([NoEnglish] >= 22)

- Click the **Select From Set** button. *Caution—Don't click* **New Set**, **or you will select all of the increased-risk areas in the city rather than just those in the flooded areas.**
- Open the **NYC Census Data** attribute table, and use the **Field ▸ Statistics** function to find the total number of people living in the increased-risk census block groups within the flooded areas.

3. What is the total number of people in the selected census block groups that meet one or more of the increased-risk criteria?

4. What percentage of the total baseline population does this represent?

    % of total population (Q3/Q1 × 100) = _____

Important query note

If the query selects more census block groups than you started with, or no census block groups at all, you have made an error. If this happens, redo the entire process beginning with the Select By Theme operation on page 70.

5. What percentage of the population in the Category 4 storm surge area is at increased risk?

% of increased-risk population in
the Category 4 storm surge area (Q3/Q2 × 100) = _____

To break down this total by category, you will find the average percentage of each increased-risk group and multiply it by the total affected population.

▶ Scroll the attribute table across and click the *Elderly* field name to highlight it. Choose **Field ▶ Statistics**, and read the **Mean** value in the statistics window. This is the average percentage of elderly people living in the affected areas.

6. Record the mean percentage of the elderly population in the table below. Repeat this process to find the mean percentage for the *Poor* and *NoEnglish* populations and record them in the table.

Calculating the affected population

Example: Non-English speakers make up 12.4% of a total affected population of 6,000,000. (Your numbers will be different.) The affected population of non-English speakers is calculated as:

0.124 × 6,000,000 = 744,000

Increased-risk category	Mean %	Total affected population (from question 3)	Affected population in category (Mean % × Total affected population)
Elderly			
Poor			
NoEnglish			

7. Enter the total affected population from question 3 in the table and use it to calculate the affected population for each increased-risk category.

▶ When you are finished, close the Attributes of NYC Census Data window.

On the map, the census block groups containing high concentrations of increased-risk population are highlighted yellow.

8. Looking at the map, are the concentrations of increased-risk groups evenly dispersed or are they clustered? What do you think accounts for the distribution you observe?

9. How might the distribution of increased-risk groups help or hinder efforts to warn the citizens of the approaching hurricane?

10. Based on what you've learned in this investigation, how would you characterize the impact of a hurricane on New York City. Would it be minor, major, or catastrophic? Explain your answer.

Activity 4.4

Assessing infrastructure

Investigating storm surge and population demographics of New York City is just one of many possible scenarios worth exploring in a study of the effects of a major hurricane on New York City. Natural disasters affect people, and the societies they are part of, in many different ways. The previous example looked at the potential direct impact of hurricane floodwaters on high-risk populations in New York City. The people living in the Category 4 storm surge areas would be physically in danger of drowning or being struck by floating debris. What are some of the other hazards New Yorkers face?

Other hazards from hurricanes

Wind damage

Another example of a hurricane's direct impact is severe wind damage to mobile home parks or areas where housing is more than fifty years old. These structures can be more susceptible to severe wind damage (i.e., collapse or disintegration) that would directly endanger their inhabitants. An emergency planner would certainly want to know where such areas were located, who lived there, and how they might warn or evacuate the residents before the disaster strikes.

Indirect effects

In addition to direct effects from natural disasters, there are indirect effects, in which the natural disaster damages or destroys some part of the infrastructure that supports modern society. We all take clean water, electricity, and food for granted, assuming that it will always be easily available. While few people are killed or injured by the actual event, many can be inconvenienced or become ill if they don't have safe drinking water, electricity, or food for an extended period of time.

Extended investigations

The following pages describe other investigations that you can carry out using the data included with this project. Each investigation describes the themes available for examining different potential impacts of a major hurricane on New York City.

Working in teams, read through the descriptions of the investigations and choose one to work on as a group. Plan and carry out the investigation and be prepared to share your findings with the class. Your instructor may provide you with additional research and presentation guidelines.

Be creative!
Now that you know how to use the GIS, this is your opportunity to be creative. See what else you can discover by experimenting and playing with the data. Have fun, but be sure to clean up New York City by turning off the storm surge when you're done!

Extended Investigation 1

Shelters and hospitals

In planning for disasters, it is important for emergency planners to determine the capacity of a city or town to take care of people injured or displaced from their homes. This view allows you to use ArcView to look at the number of hospital beds and shelters that might be lost due to hurricane-related flooding and to draw conclusions about New York City's ability to care for people in need during a natural disaster.

Investigation questions
- How many hospital beds might be lost due to flooding?
- How many shelters (schools) would be flooded and not available as temporary housing?
- How many hospital beds and shelters are outside the flood areas?
- Where would you transport people to safe havens or hospitals outside the flood areas?

Themes available
(Use the **NYC Shelter and Hospital Assessment** view.)
- **NYC Hospitals** – New York City hospital locations
- **NYC Schools** – NYC public schools (could be used as storm shelters)

Extended Investigation 2

Wind hazards

Manhattan, the heart of New York City, has some of the tallest buildings in the world. Many of these skyscrapers are glass towers that rise 70 to 100 stories above the city streets. At these altitudes, the wind gusts could be even stronger than at ground level, meaning that Category 4 hurricane winds could be well over 130 miles per hour.

Exactly how well a skyscraper would stand up to such winds isn't known, but chances are good that windows would shatter and glass would be a major component of the flying debris.

This investigation examines the risk from skyscraper window shattering, by defining 100-meter buffer zones surrounding the tall buildings. The number of people at risk from flying debris can then be determined by using the census data with the skyscraper buffers in a Select By Theme operation. This number is low because the census only where people live, not where they work. Some of these large buildings can hold tens of thousands of employees.

Investigation topics

- Identify the census block groups that intersect the Skyscraper Hazard Zones. Find the population density for those census block groups, determine the mean population density, and apply that population density to the area of the NYC Skyscraper Hazard Zones to estimate the affected population.

Themes available

(Use the **NYC Skyscraper Wind Hazard Assessment** view.)

- **Skyscrapers** – Building outlines of the 30 tallest buildings in New York City.
- **Skyscraper Hazard Zones** – 100-meter buffer zones surrounding the building perimeters that could be hazardous due to windows blowing out of the tallest buildings.
- **Census Data**

Extended Investigation 3

Superfund sites

A serious hazard in any flood event is the possibility of water pollution as flood waters inundate landfills, sewage treatment facilities, and industrial sites and pick up chemical or biological contaminants. This investigation examines the locations of EPA Superfund sites in relation to the predicted hurricane storm surge and uses the GIS to identify areas or populations at risk from contaminated floodwaters.

What is a Superfund site?

Superfund sites are places that have been identified as being the worst hazardous waste sites in the nation. The Superfund Program is run by the US Environmental Protection Agency (EPA), in cooperation with local governments. Established by Congress in 1980, its mission is to locate, investigate, and clean up these sites.

For more information, visit the Superfund website at

http://www.epa.gov/superfund

Investigation topics

- Identify Superfund sites that would be inundated by a Category 4 storm surge. Buffer the inundated sites using the Geoprocessing Wizard Extension (available from ESRI) and perform a Select By Theme operation to determine affected population.
- Identify hazards contained in the sites and try to determine their impact on water and air quality.

Themes available

(Use the **NYC Superfund Site Hazard Assessment** view.)

- **Superfund Sites** – CERCLIS Sites from EPA Region 2 GIS repository

Extended Investigation 4

Transportation networks

Roads, railways, and airports are the circulation system of cities, allowing people and goods to flow freely. People get to work and school, raw materials get to factories, food gets to supermarkets and restaurants. Disable part of the transportation network, and society grinds to a halt.

In this investigation, you can explore the impact of a major hurricane on New York City's transportation infrastructure by looking at the railroad lines and roads that cross areas inundated by storm surge flood water. In addition, you can add data to the view to examine evacuation routes and accessibility to shelters and hospitals.

Investigation topics
- Determine the length of roads and railroads flooded by a Category 4 storm surge.
- Identify evacuation routes open or closed due to storm flooding.

Themes available
(Use the **NYC Transportation Network Impact Assessment** view.)
- **Road Network** – New York City interstate highways and major roads.
- **Railroads** – New York City railroads.